U•X•L Encyclopedia of Science

U·X·L Encyclopedia of Science

Second Edition

Volume 2: At-Car

Rob Nagel, Editor

U·X·L

GALE GROUP

THOMSON LEARNING™

Detroit • New York • San Diego • San Francisco
Boston • New Haven, Conn. • Waterville, Maine
London • Munich

U•X•L Encyclopedia of Science Second Edition

Rob Nagel, *Editor*

Staff

Elizabeth Shaw Grunow, *U•X•L Editor*

Julie Carnagie, *Contributing Editor*

Carol DeKane Nagel, *U•X•L Managing Editor*

Thomas L. Romig, *U•X•L Publisher*

Shalice Shah-Caldwell, *Permissions Associate (Pictures)*

Robyn Young, *Imaging and Multimedia Content Editor*

Rita Wimberley, *Senior Buyer*

Pamela A. E. Galbreath, *Senior Art Designer*

Michelle Cadorée, *Indexing*

GGS Information Services, *Typesetting*

On the front cover: Nikola Tesla with one of his generators, reproduced by permission of the Granger Collection.

On the back cover: The flow of red blood cells through blood vessels, reproduced by permission of Phototake.

Library of Congress Cataloging-in-Publication Data

U-X-L encyclopedia of science.—2nd ed. / Rob Nagel, editor
p.cm.
Includes bibliographical references and indexes.
Contents: v.1. A-As — v.2. At-Car — v.3. Cat-Cy — v.4. D-Em — v.5. En-G — v.6. H-Mar — v.7. Mas-O — v.8. P-Ra — v.9. Re-St — v.10. Su-Z.
Summary: Includes 600 topics in the life, earth, and physical sciences as well as in engineering, technology, math, environmental science, and psychology.
ISBN 0-7876-5432-9 (set : acid-free paper) — ISBN 0-7876-5433-7 (v.1 : acid-free paper) — ISBN 0-7876-5434-5 (v.2 : acid-free paper) — ISBN 0-7876-5435-3 (v.3 : acid-free paper) — ISBN 0-7876-5436-1 (v.4 : acid-free paper) — ISBN 0-7876-5437-X (v.5 : acid-free paper) — ISBN 0-7876-5438-8 (v.6 : acid-free paper) — ISBN 0-7876-5439-6 (v.7 : acid-free paper) — ISBN 0-7876-5440-X (v.8 : acid-free paper) — ISBN 0-7876-5441-8 (v.9 : acid-free paper) — ISBN 0-7876-5775-1 (v.10 : acid-free paper)

1. Science-Encyclopedias, Juvenile. 2. Technology-Encyclopedias, Juvenile. [1. Science-Encyclopedias. 2. Technology-Encyclopedias.] I. Title: UXL encyclopedia of science. II. Nagel, Rob.

Q121.U18 2001
503-dc21

2001035562

Printed in the United States of America

10 9 8 7 6 5 4 3 2 1

Table of Contents

Reader's Guide vii

Entries by Scientific Field ix

Volume 1: A-As 1
Where to Learn More xxxi
Index xxxv

Volume 2: At-Car 211
Where to Learn More xxxi
Index xxxv

Volume 3: Cat-Cy 413
Where to Learn More xxxi
Index xxxv

Volume 4: D-Em 611
Where to Learn More xxxi
Index xxxv

Volume 5: En-G 793
Where to Learn More xxxi
Index xxxv

Volume 6: H-Mar 1027
Where to Learn More xxxi
Index xxxv

Volume 7: Mas-O 1235
Where to Learn More xxxi
Index xxxv

Contents

Volume 8: P-Ra 1457
Where to Learn More xxxi
Index xxxv

Volume 9: Re-St 1647
Where to Learn More xxxi
Index xxxv

Volume 10: Su-Z 1829
Where to Learn More xxxi
Index xxxv

Reader's Guide

Demystify scientific theories, controversies, discoveries, and phenomena with the *U•X•L Encyclopedia of Science,* Second Edition.

This alphabetically organized ten-volume set opens up the entire world of science in clear, nontechnical language. More than 600 entries—an increase of more than 10 percent from the first edition—provide fascinating facts covering the entire spectrum of science. This second edition features more than 50 new entries and more than 100 updated entries. These informative essays range from 250 to 2,500 words, many of which include helpful sidebar boxes that highlight fascinating facts and phenomena. Topics profiled are related to the physical, life, and earth sciences, as well as to math, psychology, engineering, technology, and the environment.

In addition to solid information, the *Encyclopedia* also provides these features:

- "Words to Know" boxes that define commonly used terms
- Extensive cross references that lead directly to related entries
- A table of contents by scientific field that organizes the entries
- More than 600 color and black-and-white photos and technical drawings
- Sources for further study, including books, magazines, and Web sites

Each volume concludes with a cumulative subject index, making it easy to locate quickly the theories, people, objects, and inventions discussed throughout the *U•X•L Encyclopedia of Science,* Second Edition.

Suggestions

We welcome any comments on this work and suggestions for entries to feature in future editions of *U•X•L Encyclopedia of Science.* Please write: Editors, *U•X•L Encyclopedia of Science,* U•X•L, Gale Group, 27500 Drake Road, Farmington Hills, Michigan, 48331-3535; call toll-free: 800-877-4253; fax to: 248-699-8097; or send an e-mail via www.galegroup.com.

Entries by Scientific Field

Boldface indicates volume numbers.

Acoustics

Acoustics **1**:17
Compact disc **3**:531
Diffraction **4**:648
Echolocation **4**:720
Magnetic recording/ audiocassette **6**:1209
Sonar **9**:1770
Ultrasonics **10**:1941
Video recording **10**:1968

Aerodynamics

Aerodynamics **1**:39
Fluid dynamics **5**:882

Aeronautical engineering

Aircraft **1**:74
Atmosphere observation **2**:215
Balloon **1**:261
Jet engine **6**:1143
Rockets and missiles **9**:1693

Aerospace engineering

International Ultraviolet Explorer **6**:1120
Rockets and missiles **9**:1693
Satellite **9**:1707
Spacecraft, manned **9**:1777
Space probe **9**:1783
Space station, international **9**:1788
Telescope **10**:1869

Agriculture

Agriculture **1**:62
Agrochemical **1**:65
Aquaculture **1**:166
Biotechnology **2**:309
Cotton **3**:577
Crops **3**:582
DDT (dichlorodiphenyl-trichloroethane) **4**:619
Drift net **4**:680
Forestry **5**:901
Genetic engineering **5**:973
Organic farming **7**:1431
Slash-and-burn agriculture **9**:1743
Soil **9**:1758

Anatomy and physiology

Anatomy **1**:138
Blood **2**:326

Brain 2:337
Cholesterol 3:469
Chromosome 3:472
Circulatory system 3:480
Digestive system 4:653
Ear 4:693
Endocrine system 5:796
Excretory system 5:839
Eye 5:848
Heart 6:1037
Human Genome Project 6:1060
Immune system 6:1082
Integumentary system 6:1109
Lymphatic system 6:1198
Muscular system 7:1309
Nervous system 7:1333
Physiology 8:1516
Reproductive system 9:1667
Respiratory system 9:1677
Skeletal system 9:1739
Smell 9:1750
Speech 9:1796
Taste 10:1861
Touch 10:1903

Anesthesiology

Alternative medicine 1:118
Anesthesia 1:142

Animal husbandry

Agrochemical 1:65
Biotechnology 2:309
Crops 3:582
Genetic engineering 5:973
Organic farming 7:1431

Anthropology

Archaeoastronomy 1:171
Dating techniques 4:616
Forensic science 5:898
Gerontology 5:999
Human evolution 6:1054
Mounds, earthen 7:1298
Petroglyphs and pictographs 8:1491

Aquaculture

Aquaculture 1:166
Crops 3:582
Drift net 4:680
Fish 5:875

Archaeology

Archaeoastronomy 1:171
Archaeology 1:173
Dating techniques 4:616
Fossil and fossilization 5:917
Half-life 6:1027
Nautical archaeology 7:1323
Petroglyphs and pictographs 8:1491

Artificial intelligence

Artificial intelligence 1:188
Automation 2:242

Astronomy

Archaeoastronomy 1:171
Asteroid 1:200
Astrophysics 1:207
Big bang theory 2:273
Binary star 2:276
Black hole 2:322
Brown dwarf 2:358
Calendar 2:372
Celestial mechanics 3:423
Comet 3:527
Constellation 3:558
Cosmic ray 3:571

Cosmology 3:574
Dark matter 4:613
Earth (planet) 4:698
Eclipse 4:723
Extrasolar planet 5:847
Galaxy 5:941
Gamma ray 5:949
Gamma-ray burst 5:952
Gravity and gravitation 5:1012
Infrared astronomy 6:1100
International Ultraviolet Explorer 6:1120
Interstellar matter 6:1130
Jupiter (planet) 6:1146
Light-year 6:1190
Mars (planet) 6:1228
Mercury (planet) 7:1250
Meteor and meteorite 7:1262
Moon 7:1294
Nebula 7:1327
Neptune (planet) 7:1330
Neutron star 7:1339
Nova 7:1359
Orbit 7:1426
Pluto (planet) 8:1539
Quasar 8:1609
Radio astronomy 8:1633
Red giant 9:1653
Redshift 9:1654
Satellite 9:1707
Saturn (planet) 9:1708
Seasons 9:1726
Solar system 9:1762
Space 9:1776
Spacecraft, manned 9:1777
Space probe 9:1783
Space station, international 9:1788
Star 9:1801
Starburst galaxy 9:1806
Star cluster 9:1808
Stellar magnetic fields 9:1820
Sun 10:1844
Supernova 10:1852
Telescope 10:1869
Ultraviolet astronomy 10:1943
Uranus (planet) 10:1952
Variable star 10:1963
Venus (planet) 10:1964
White dwarf 10:2027
X-ray astronomy 10:2038

Astrophysics

Astrophysics 1:207
Big bang theory 2:273
Binary star 2:276
Black hole 2:322
Brown dwarf 2:358
Celestial mechanics 3:423
Cosmic ray 3:571
Cosmology 3:574
Dark matter 4:613
Galaxy 5:941
Gamma ray 5:949
Gamma-ray burst 5:952
Gravity and gravitation 5:1012
Infrared astronomy 6:1100
International Ultraviolet Explorer 6:1120
Interstellar matter 6:1130
Light-year 6:1190
Neutron star 7:1339
Orbit 7:1426
Quasar 8:1609
Radio astronomy 8:1633
Red giant 9:1653
Redshift 9:1654
Space 9:1776
Star 9:1801
Starburst galaxy 9:1806
Star cluster 9:1808
Stellar magnetic fields 9:1820
Sun 10:1844

Supernova	**10:**1852
Ultraviolet astronomy	**10:**1943
Uranus (planet)	**10:**1952
Variable star	**10:**1963
White dwarf	**10:**2027
X-ray astronomy	**10:**2038

Atomic/Nuclear physics

Actinides	**1:**23
Alkali metals	**1:**99
Alkali earth metals	**1:**102
Alternative energy sources	**1:**111
Antiparticle	**1:**163
Atom	**2:**226
Atomic mass	**2:**229
Atomic theory	**2:**232
Chemical bond	**3:**453
Dating techniques	**4:**616
Electron	**4:**768
Half-life	**6:**1027
Ionization	**6:**1135
Isotope	**6:**1141
Lanthanides	**6:**1163
Mole (measurement)	**7:**1282
Molecule	**7:**1285
Neutron	**7:**1337
Noble gases	**7:**1349
Nuclear fission	**7:**1361
Nuclear fusion	**7:**1366
Nuclear medicine	**7:**1372
Nuclear power	**7:**1374
Nuclear weapons	**7:**1381
Particle accelerators	**8:**1475
Quantum mechanics	**8:**1607
Radiation	**8:**1619
Radiation exposure	**8:**1621
Radiology	**8:**1637
Subatomic particles	**10:**1829
X ray	**10:**2033

Automotive engineering

Automobile	**2:**245
Diesel engine	**4:**646
Internal-combustion engine	**6:**1117

Bacteriology

Bacteria	**2:**253
Biological warfare	**2:**287
Disease	**4:**669
Legionnaire's disease	**6:**1179

Ballistics

Ballistics	**2:**260
Nuclear weapons	**7:**1381
Rockets and missiles	**9:**1693

Biochemistry

Amino acid	**1:**130
Biochemistry	**2:**279
Carbohydrate	**2:**387
Cell	**3:**428
Cholesterol	**3:**469
Enzyme	**5:**812
Fermentation	**5:**864
Hormones	**6:**1050
Human Genome Project	**6:**1060
Lipids	**6:**1191
Metabolism	**7:**1255
Nucleic acid	**7:**1387
Osmosis	**7:**1436
Photosynthesis	**8:**1505
Proteins	**8:**1586
Respiration	**9:**1672
Vitamin	**10:**1981
Yeast	**10:**2043

Biology

Adaptation 1:26
Algae 1:91
Amino acid 1:130
Amoeba 1:131
Amphibians 1:134
Anatomy 1:138
Animal 1:145
Antibody and antigen 1:159
Arachnids 1:168
Arthropods 1:183
Bacteria 2:253
Behavior 2:270
Biochemistry 2:279
Biodegradable 2:280
Biodiversity 2:281
Biological warfare 2:287
Biology 2:290
Biome 2:293
Biophysics 2:302
Biosphere 2:304
Biotechnology 2:309
Birds 2:312
Birth 2:315
Birth defects 2:319
Blood 2:326
Botany 2:334
Brain 2:337
Butterflies 2:364
Canines 2:382
Carbohydrate 2:387
Carcinogen 2:406
Cell 3:428
Cellulose 3:442
Cetaceans 3:448
Cholesterol 3:469
Chromosome 3:472
Circulatory system 3:480
Clone and cloning 3:484
Cockroaches 3:505
Coelacanth 3:508
Contraception 3:562
Coral 3:566
Crustaceans 3:590
Cryobiology 3:593
Digestive system 4:653
Dinosaur 4:658
Disease 4:669
Ear 4:693
Embryo and embryonic development 4:785
Endocrine system 5:796
Enzyme 5:812
Eutrophication 5:828
Evolution 5:832
Excretory system 5:839
Eye 5:848
Felines 5:855
Fermentation 5:864
Fertilization 5:867
Fish 5:875
Flower 5:878
Forestry 5:901
Forests 5:907
Fungi 5:930
Genetic disorders 5:966
Genetic engineering 5:973
Genetics 5:980
Heart 6:1037
Hibernation 6:1046
Hormones 6:1050
Horticulture 6:1053
Human Genome Project 6:1060
Human evolution 6:1054
Immune system 6:1082
Indicator species 6:1090
Insects 6:1103
Integumentary system 6:1109
Invertebrates 6:1133
Kangaroos and wallabies 6:1153
Leaf 6:1172
Lipids 6:1191
Lymphatic system 6:1198
Mammals 6:1222

Mendelian laws of inheritance 7:1246
Metabolism 7:1255
Metamorphosis 7:1259
Migration (animals) 7:1271
Molecular biology 7:1283
Mollusks 7:1288
Muscular system 7:1309
Mutation 7:1314
Nervous system 7:1333
Nucleic acid 7:1387
Osmosis 7:1436
Parasites 8:1467
Photosynthesis 8:1505
Phototropism 8:1508
Physiology 8:1516
Plague 8:1518
Plankton 8:1520
Plant 8:1522
Primates 8:1571
Proteins 8:1586
Protozoa 8:1590
Puberty 8:1599
Rain forest 8:1641
Reproduction 9:1664
Reproductive system 9:1667
Reptiles 9:1670
Respiration 9:1672
Respiratory system 9:1677
Rh factor 9:1683
Seed 9:1729
Sexually transmitted diseases 9:1735
Skeletal system 9:1739
Smell 9:1750
Snakes 9:1752
Speech 9:1796
Sponges 9:1799
Taste 10:1861
Touch 10:1903
Tree 10:1927
Tumor 10:1934
Vaccine 10:1957
Vertebrates 10:1967
Virus 10:1974
Vitamin 10:1981
Wetlands 10:2024
Yeast 10:2043

Biomedical engineering

Electrocardiogram 4:751
Radiology 8:1637

Biotechnology

Biotechnology 2:309
Brewing 2:352
Fermentation 5:864
Vaccine 10:1957

Botany

Botany 2:334
Cellulose 3:442
Cocaine 3:501
Cotton 3:577
Flower 5:878
Forestry 5:901
Forests 5:907
Horticulture 6:1053
Leaf 6:1172
Marijuana 6:1224
Photosynthesis 8:1505
Phototropism 8:1508
Plant 8:1522
Seed 9:1729
Tree 10:1927

Cartography

Cartography 2:410
Geologic map 5:986

Cellular biology

Amino acid 1:130
Carbohydrate 2:387
Cell 3:428
Cholesterol 3:469
Chromosome 3:472
Genetics 5:980
Lipids 6:1191
Osmosis 7:1436
Proteins 8:1586

Chemistry

Acids and bases 1:14
Actinides 1:23
Aerosols 1:43
Agent Orange 1:54
Agrochemical 1:65
Alchemy 1:82
Alcohols 1:88
Alkali metals 1:99
Alkaline earth metals 1:102
Aluminum family 1:122
Atom 2:226
Atomic mass 2:229
Atomic theory 2:232
Biochemistry 2:279
Carbon dioxide 2:393
Carbon family 2:395
Carbon monoxide 2:403
Catalyst and catalysis 2:413
Chemical bond 3:453
Chemical w\arfare 3:457
Chemistry 3:463
Colloid 3:515
Combustion 3:522
Composite materials 3:536
Compound, chemical 3:541
Crystal 3:601
Cyclamate 3:608
DDT (dichlorodiphenyl-trichloroethane) 4:619
Diffusion 4:651
Dioxin 4:667
Distillation 4:675
Dyes and pigments 4:686
Electrolysis 4:755
Element, chemical 4:774
Enzyme 5:812
Equation, chemical 5:815
Equilibrium, chemical 5:817
Explosives 5:843
Fermentation 5:864
Filtration 5:872
Formula, chemical 5:914
Halogens 6:1030
Hormones 6:1050
Hydrogen 6:1068
Industrial minerals 6:1092
Ionization 6:1135
Isotope 6:1141
Lanthanides 6:1163
Lipids 6:1191
Metabolism 7:1255
Mole (measurement) 7:1282
Molecule 7:1285
Nitrogen family 7:1344
Noble gases 7:1349
Nucleic acid 7:1387
Osmosis 7:1436
Oxidation-reduction reaction 7:1439
Oxygen family 7:1442
Ozone 7:1450
Periodic table 8:1486
pH 8:1495
Photochemistry 8:1498
Photosynthesis 8:1505
Plastics 8:1532
Poisons and toxins 8:1542
Polymer 8:1563
Proteins 8:1586
Qualitative analysis 8:1603
Quantitative analysis 8:1604

Reaction, chemical 9:1647
Respiration 9:1672
Soaps and detergents 9:1756
Solution 9:1767
Transition elements 10:1913
Vitamin 10:1981
Yeast 10:2043

Civil engineering

Bridges 2:354
Canal 2:376
Dam 4:611
Lock 6:1192

Climatology

Global climate 5:1006
Ice ages 6:1075
Seasons 9:1726

Communications/Graphic arts

Antenna 1:153
CAD/CAM 2:369
Cellular/digital technology 3:439
Compact disc 3:531
Computer software 3:549
DVD technology 4:684
Hologram and holography 6:1048
Internet 6:1123
Magnetic recording/audiocassette 6:1209
Microwave communication 7:1268
Petroglyphs and pictographs 8:1491
Photocopying 8:1499
Radio 8:1626
Satellite 9:1707
Telegraph 10:1863
Telephone 10:1866
Television 10:1875
Video recording 10:1968

Computer science

Artificial intelligence 1:188
Automation 2:242
CAD/CAM 2:369
Calculator 2:370
Cellular/digital technology 3:439
Compact disc 3:531
Computer, analog 3:546
Computer, digital 3:547
Computer software 3:549
Internet 6:1123
Mass production 7:1236
Robotics 9:1690
Virtual reality 10:1969

Cosmology

Astrophysics 1:207
Big Bang theory 2:273
Cosmology 3:574
Galaxy 5:941
Space 9:1776

Cryogenics

Cryobiology 3:593
Cryogenics 3:595

Dentistry

Dentistry 4:626
Fluoridation 5:889

Ecology/Environmental science

Acid rain 1:9
Alternative energy sources 1:111
Biodegradable 2:280
Biodiversity 2:281

Bioenergy 2:284
Biome 2:293
Biosphere 2:304
Carbon cycle 2:389
Composting 3:539
DDT (dichlorodiphenyl-trichloroethane) 4:619
Desert 4:634
Dioxin 4:667
Drift net 4:680
Drought 4:682
Ecology 4:725
Ecosystem 4:728
Endangered species 5:793
Environmental ethics 5:807
Erosion 5:820
Eutrophication 5:828
Food web and food chain 5:894
Forestry 5:901
Forests 5:907
Gaia hypothesis 5:935
Greenhouse effect 5:1016
Hydrologic cycle 6:1071
Indicator species 6:1090
Nitrogen cycle 7:1342
Oil spills 7:1422
Organic farming 7:1431
Paleoecology 8:1457
Pollution 8:1549
Pollution control 8:1558
Rain forest 8:1641
Recycling 9:1650
Succession 10:1837
Waste management 10:2003
Wetlands 10:2024

Electrical engineering

Antenna 1:153
Battery 2:268
Cathode 3:415
Cathode-ray tube 3:417
Cell, electrochemical 3:436
Compact disc 3:531
Diode 4:665
Electric arc 4:734
Electric current 4:737
Electricity 4:741
Electric motor 4:747
Electrocardiogram 4:751
Electromagnetic field 4:758
Electromagnetic induction 4:760
Electromagnetism 4:766
Electronics 4:773
Fluorescent light 5:886
Generator 5:962
Incandescent light 6:1087
Integrated circuit 6:1106
LED (light-emitting diode) 6: 1176
Magnetic recording/audiocassette 6:1209
Radar 8:1613
Radio 8:1626
Superconductor 10:1849
Telegraph 10:1863
Telephone 10:1866
Television 10:1875
Transformer 10:1908
Transistor 10:1910
Ultrasonics 10:1941
Video recording 10:1968

Electronics

Antenna 1:153
Battery 2:268
Cathode 3:415
Cathode-ray tube 3:417
Cell, electrochemical 3:436
Compact disc 3:531
Diode 4:665
Electric arc 4:734
Electric current 4:737
Electricity 4:741
Electric motor 4:747

Electromagnetic field 4:758
Electromagnetic induction 4:760
Electronics 4:773
Generator 5:962
Integrated circuit 6:1106
LED (light-emitting diode) 6:1176
Magnetic recording/ audiocassette 6:1209
Radar 8:1613
Radio 8:1626
Superconductor 10:1849
Telephone 10:1866
Television 10:1875
Transformer 10:1908
Transistor 10:1910
Ultrasonics 10:1941
Video recording 10:1968

Embryology

Embryo and embryonic development 4:785
Fertilization 5:867
Reproduction 9:1664
Reproductive system 9:1667

Engineering

Aerodynamics 1:39
Aircraft 1:74
Antenna 1:153
Automation 2:242
Automobile 2:245
Balloon 1:261
Battery 2:268
Bridges 2:354
Canal 2:376
Cathode 3:415
Cathode-ray tube 3:417
Cell, electrochemical 3:436
Compact disc 3:531
Dam 4:611
Diesel engine 4:646
Diode 4:665
Electric arc 4:734
Electric current 4:737
Electric motor 4:747
Electricity 4:741
Electrocardiogram 4:751
Electromagnetic field 4:758
Electromagnetic induction 4:760
Electromagnetism 4:766
Electronics 4:773
Engineering 5:805
Fluorescent light 5:886
Generator 5:962
Incandescent light 6:1087
Integrated circuit 6:1106
Internal-combustion engine 6:1117
Jet engine 6:1143
LED (light-emitting diode) 6: 1176
Lock 6:1192
Machines, simple 6:1203
Magnetic recording/ audiocassette 6:1209
Mass production 7:1236
Radar 8:1613
Radio 8:1626
Steam engine 9:1817
Submarine 10:1834
Superconductor 10:1849
Telegraph 10:1863
Telephone 10:1866
Television 10:1875
Transformer 10:1908
Transistor 10:1910
Ultrasonics 10:1941
Video recording 10:1968

Entomology

Arachnids 1:168
Arthropods 1:183

Butterflies **2:**364
Cockroaches **3:**505
Insects **6:**1103
Invertebrates **6:**1133
Metamorphosis **7:**1259

Epidemiology

Biological warfare **2:**287
Disease **4:**669
Ebola virus **4:**717
Plague **8:**1518
Poliomyelitis **8:**1546
Sexually transmitted diseases **9:**1735
Vaccine **10:**1957

Evolutionary biology

Adaptation **1:**26
Evolution **5:**832
Human evolution **6:**1054
Mendelian laws of inheritance **7:**1246

Food science

Brewing **2:**352
Cyclamate **3:**608
Food preservation **5:**890
Nutrition **7:**1399

Forensic science

Forensic science **5:**898

Forestry

Forestry **5:**901
Forests **5:**907
Rain forest **8:**1641
Tree **10:**1927

General science

Alchemy **1:**82
Chaos theory **3:**451
Metric system **7:**1265
Scientific method **9:**1722
Units and standards **10:**1948

Genetic engineering

Biological warfare **2:**287
Biotechnology **2:**309
Clone and cloning **3:**484
Genetic engineering **5:**973

Genetics

Biotechnology **2:**309
Birth defects **2:**319
Cancer **2:**379
Carcinogen **2:**406
Chromosome **3:**472
Clone and cloning **3:**484
Genetic disorders **5:**966
Genetic engineering **5:**973
Genetics **5:**980
Human Genome Project **6:**1060
Mendelian laws of inheritance **7:**1246
Mutation **7:**1314
Nucleic acid **7:**1387

Geochemistry

Coal **3:**492
Earth (planet) **4:**698
Earth science **4:**707
Earth's interior **4:**708
Glacier **5:**1000
Minerals **7:**1273
Rocks **9:**1701
Soil **9:**1758

Geography

Africa **1:**49
Antarctica **1:**147
Asia **1:**194
Australia **2:**238
Biome **2:**293
Cartography **2:**410
Coast and beach **3:**498
Desert **4:**634
Europe **5:**823
Geologic map **5:**986
Island **6:**1137
Lake **6:**1159
Mountain **7:**1301
North America **7:**1352
River **9:**1685
South America **9:**1772

Geology

Catastrophism **3:**415
Cave **3:**420
Coal **3:**492
Coast and beach **3:**498
Continental margin **3:**560
Dating techniques **4:**616
Desert **4:**634
Earthquake **4:**702
Earth science **4:**707
Earth's interior **4:**708
Erosion **5:**820
Fault **5:**855
Geologic map **5:**986
Geologic time **5:**990
Geology **5:**993
Glacier **5:**1000
Hydrologic cycle **6:**1071
Ice ages **6:**1075
Iceberg **6:**1078
Industrial minerals **6:**1092
Island **6:**1137
Lake **6:**1159
Minerals **7:**1273
Mining **7:**1278
Mountain **7:**1301
Natural gas **7:**1319
Oil drilling **7:**1418
Oil spills **7:**1422
Petroleum **8:**1492
Plate tectonics **8:**1534
River **9:**1685
Rocks **9:**1701
Soil **9:**1758
Uniformitarianism **10:**1946
Volcano **10:**1992
Water **10:**2010

Geophysics

Earth (planet) **4:**698
Earth science **4:**707
Fault **5:**855
Plate tectonics **8:**1534

Gerontology

Aging and death **1:**59
Alzheimer's disease **1:**126
Arthritis **1:**181
Dementia **4:**622
Gerontology **5:**999

Gynecology

Contraception **3:**562
Fertilization **5:**867
Gynecology **5:**1022
Puberty **8:**1599
Reproduction **9:**1664

Health/Medicine

Acetylsalicylic acid **1:**6
Addiction **1:**32
Attention-deficit hyperactivity disorder (ADHD) **2:**237

Depression 4:630
AIDS (acquired immunodeficiency syndrome) 1:70
Alcoholism 1:85
Allergy 1:106
Alternative medicine 1:118
Alzheimer's disease 1:126
Amino acid 1:130
Anesthesia 1:142
Antibiotics 1:155
Antiseptics 1:164
Arthritis 1:181
Asthma 1:204
Attention-deficit hyperactivity disorder (ADHD) 2:237
Birth defects 2:319
Blood supply 2:330
Burn 2:361
Carcinogen 2:406
Carpal tunnel syndrome 2:408
Cholesterol 3:469
Cigarette smoke 3:476
Cocaine 3:501
Contraception 3:562
Dementia 4:622
Dentistry 4:626
Depression 4:630
Diabetes mellitus 4:638
Diagnosis 4:640
Dialysis 4:644
Disease 4:669
Dyslexia 4:690
Eating disorders 4:711
Ebola virus 4:717
Electrocardiogram 4:751
Fluoridation 5:889
Food preservation 5:890
Genetic disorders 5:966
Genetic engineering 5:973
Genetics 5:980
Gerontology 5:999
Gynecology 5:1022
Hallucinogens 6:1027
Immune system 6:1082
Legionnaire's disease 6:1179
Lipids 6:1191
Malnutrition 6:1216
Marijuana 6:1224
Multiple personality disorder 7:1305
Nuclear medicine 7:1372
Nutrition 7:1399
Obsession 7:1405
Orthopedics 7:1434
Parasites 8:1467
Phobia 8:1497
Physical therapy 8:1511
Plague 8:1518
Plastic surgery 8:1527
Poliomyelitis 8:1546
Prosthetics 8:1579
Protease inhibitor 8:1583
Psychiatry 8:1592
Psychology 8:1594
Psychosis 8:1596
Puberty 8:1599
Radial keratotomy 8:1615
Radiology 8:1637
Rh factor 9:1683
Schizophrenia 9:1716
Sexually transmitted diseases 9:1735
Sleep and sleep disorders 9:1745
Stress 9:1826
Sudden infant death syndrome (SIDS) 10:1840
Surgery 10:1855
Tranquilizers 10:1905
Transplant, surgical 10:1923
Tumor 10:1934
Vaccine 10:1957
Virus 10:1974
Vitamin 10:1981
Vivisection 10:1989

Horticulture

Horticulture **6:**1053
Plant **8:**1522
Seed **9:**1729
Tree **10:**1927

Immunology

Allergy **1:**106
Antibiotics **1:**155
Antibody and antigen **1:**159
Immune system **6:**1082
Vaccine **10:**1957

Marine biology

Algae **1:**91
Amphibians **1:**134
Cetaceans **3:**448
Coral **3:**566
Crustaceans **3:**590
Endangered species **5:**793
Fish **5:**875
Mammals **6:**1222
Mollusks **7:**1288
Ocean zones **7:**1414
Plankton **8:**1520
Sponges **9:**1799
Vertebrates **10:**1967

Materials science

Abrasives **1:**2
Adhesives **1:**37
Aerosols **1:**43
Alcohols **1:**88
Alkaline earth metals **1:**102
Alloy **1:**110
Aluminum family **1:**122
Artificial fibers **1:**186
Asbestos **1:**191
Biodegradable **2:**280
Carbon family **2:**395
Ceramic **3:**447
Composite materials **3:**536
Dyes and pigments **4:**686
Electrical conductivity **4:**731
Electrolysis **4:**755
Expansion, thermal **5:**842
Fiber optics **5:**870
Glass **5:**1004
Halogens **6:**1030
Hand tools **6:**1036
Hydrogen **6:**1068
Industrial minerals **6:**1092
Minerals **7:**1273
Nitrogen family **7:**1344
Oxygen family **7:**1442
Plastics **8:**1532
Polymer **8:**1563
Soaps and detergents **9:**1756
Superconductor **10:**1849
Transition elements **10:**1913

Mathematics

Abacus **1:**1
Algebra **1:**97
Arithmetic **1:**177
Boolean algebra **2:**333
Calculus **2:**371
Chaos theory **3:**451
Circle **3:**478
Complex numbers **3:**534
Correlation **3:**569
Fractal **5:**921
Fraction, common **5:**923
Function **5:**927
Game theory **5:**945
Geometry **5:**995
Graphs and graphing **5:**1009
Imaginary number **6:**1081
Logarithm **6:**1195
Mathematics **7:**1241

Multiplication 7:1307
Natural numbers 7:1321
Number theory 7:1393
Numeration systems 7:1395
Polygon 8:1562
Probability theory 8:1575
Proof (mathematics) 8:1578
Pythagorean theorem 8:1601
Set theory 9:1733
Statistics 9:1810
Symbolic logic 10:1859
Topology 10:1897
Trigonometry 10:1931
Zero 10:2047

Metallurgy

Alkali metals 1:99
Alkaline earth metals 1:102
Alloy 1:110
Aluminum family 1:122
Carbon family 2:395
Composite materials 3:536
Industrial minerals 6:1092
Minerals 7:1273
Mining 7:1278
Precious metals 8:1566
Transition elements 10:1913

Meteorology

Air masses and fronts 1:80
Atmosphere, composition and structure 2:211
Atmosphere observation 2:215
Atmospheric circulation 2:218
Atmospheric optical effects 2:221
Atmospheric pressure 2:225
Barometer 2:265
Clouds 3:490
Cyclone and anticyclone 3:608
Drought 4:682
El Niño 4:782
Global climate 5:1006
Monsoon 7:1291
Ozone 7:1450
Storm surge 9:1823
Thunderstorm 10:1887
Tornado 10:1900
Weather 10:2017
Weather forecasting 10:2020
Wind 10:2028

Microbiology

Algae 1:91
Amoeba 1:131
Antiseptics 1:164
Bacteria 2:253
Biodegradable 2:280
Biological warfare 2:287
Composting 3:539
Parasites 8:1467
Plankton 8:1520
Protozoa 8:1590
Yeast 10:2043

Mineralogy

Abrasives 1:2
Ceramic 3:447
Industrial minerals 6:1092
Minerals 7:1273
Mining 7:1278

Molecular biology

Amino acid 1:130
Antibody and antigen 1:159
Biochemistry 2:279
Birth defects 2:319
Chromosome 3:472
Clone and cloning 3:484
Enzyme 5:812
Genetic disorders 5:966

Genetic engineering **5:**973
Genetics **5:**980
Hormones **6:**1050
Human Genome Project **6:**1060
Lipids **6:**1191
Molecular biology **7:**1283
Mutation **7:**1314
Nucleic acid **7:**1387
Proteins **8:**1586

Mycology

Brewing **2:**352
Fermentation **5:**864
Fungi **5:**930
Yeast **10:**2043

Nutrition

Diabetes mellitus **4:**638
Eating disorders **4:**711
Food web and food chain **5:**894
Malnutrition **6:**1216
Nutrition **7:**1399
Vitamin **10:**1981

Obstetrics

Birth **2:**315
Birth defects **2:**319
Embryo and embryonic development **4:**785

Oceanography

Continental margin **3:**560
Currents, ocean **3:**604
Ocean **7:**1407
Oceanography **7:**1411
Ocean zones **7:**1414
Tides **10:**1890

Oncology

Cancer **2:**379
Disease **4:**669
Tumor **10:**1934

Ophthalmology

Eye **5:**848
Lens **6:**1184
Radial keratotomy **8:**1615

Optics

Atmospheric optical effects **2:**221
Compact disc **3:**531
Diffraction **4:**648
Eye **5:**848
Fiber optics **5:**870
Hologram and holography **6:**1048
Laser **6:**1166
LED (light-emitting diode) **6:**1176
Lens **6:**1184
Light **6:**1185
Luminescence **6:**1196
Photochemistry **8:**1498
Photocopying **8:**1499
Telescope **10:**1869
Television **10:**1875
Video recording **10:**1968

Organic chemistry

Carbon family **2:**395
Coal **3:**492
Cyclamate **3:**608
Dioxin **4:**667
Fermentation **5:**864
Hydrogen **6:**1068
Hydrologic cycle **6:**1071
Lipids **6:**1191

Natural gas **7:**1319
Nitrogen cycle **7:**1342
Nitrogen family **7:**1344
Oil spills **7:**1422
Organic chemistry **7:**1428
Oxygen family **7:**1442
Ozone **7:**1450
Petroleum **8:**1492
Vitamin **10:**1981

Orthopedics

Arthritis **1:**181
Orthopedics **7:**1434
Prosthetics **8:**1579
Skeletal system **9:**1739

Paleontology

Dating techniques **4:**616
Dinosaur **4:**658
Evolution **5:**832
Fossil and fossilization **5:**917
Human evolution **6:**1054
Paleoecology **8:**1457
Paleontology **8:**1459

Parasitology

Amoeba **1:**131
Disease **4:**669
Fungi **5:**930
Parasites **8:**1467

Pathology

AIDS (acquired immunodeficiency syndrome) **1:**70
Alzheimer's disease **1:**126
Arthritis **1:**181
Asthma **1:**204
Attention-deficit hyperactivity disorder (ADHD) **2:**237
Bacteria **2:**253
Biological warfare **2:**287
Cancer **2:**379
Dementia **4:**622
Diabetes mellitus **4:**638
Diagnosis **4:**640
Dioxin **4:**667
Disease **4:**669
Ebola virus **4:**717
Genetic disorders **5:**966
Malnutrition **6:**1216
Orthopedics **7:**1434
Parasites **8:**1467
Plague **8:**1518
Poliomyelitis **8:**1546
Sexually transmitted diseases **9:**1735
Tumor **10:**1934
Vaccine **10:**1957
Virus **10:**1974

Pharmacology

Acetylsalicylic acid **1:**6
Antibiotics **1:**155
Antiseptics **1:**164
Cocaine **3:**501
Hallucinogens **6:**1027
Marijuana **6:**1224
Poisons and toxins **8:**1542
Tranquilizers **10:**1905

Physics

Acceleration **1:**4
Acoustics **1:**17
Aerodynamics **1:**39
Antiparticle **1:**163
Astrophysics **1:**207
Atom **2:**226
Atomic mass **2:**229
Atomic theory **2:**232
Ballistics **2:**260

Entries by Scientific Field

Battery 2:268
Biophysics 2:302
Buoyancy 2:360
Calorie 2:375
Cathode 3:415
Cathode-ray tube 3:417
Celestial mechanics 3:423
Cell, electrochemical 3:436
Chaos theory 3:451
Color 3:518
Combustion 3:522
Conservation laws 3:554
Coulomb 3:579
Cryogenics 3:595
Dating techniques 4:616
Density 4:624
Diffraction 4:648
Diode 4:665
Doppler effect 4:677
Echolocation 4:720
Elasticity 4:730
Electrical conductivity 4:731
Electric arc 4:734
Electric current 4:737
Electricity 4:741
Electric motor 4:747
Electrolysis 4:755
Electromagnetic field 4:758
Electromagnetic induction 4:760
Electromagnetic spectrum 4:763
Electromagnetism 4:766
Electron 4:768
Electronics 4:773
Energy 5:801
Evaporation 5:831
Expansion, thermal 5:842
Fiber optics 5:870
Fluid dynamics 5:882
Fluorescent light 5:886
Frequency 5:925
Friction 5:926
Gases, liquefaction of 5:955
Gases, properties of 5:959
Generator 5:962
Gravity and gravitation 5:1012
Gyroscope 5:1024
Half-life 6:1027
Heat 6:1043
Hologram and holography 6:1048
Incandescent light 6:1087
Integrated circuit 6:1106
Interference 6:1112
Interferometry 6:1114
Ionization 6:1135
Isotope 6:1141
Laser 6:1166
Laws of motion 6:1169
LED (light-emitting diode) 6:1176
Lens 6:1184
Light 6:1185
Luminescence 6:1196
Magnetic recording/audiocassette 6:1209
Magnetism 6:1212
Mass 7:1235
Mass spectrometry 7:1239
Matter, states of 7:1243
Microwave communication 7:1268
Molecule 7:1285
Momentum 7:1290
Nuclear fission 7:1361
Nuclear fusion 7:1366
Nuclear medicine 7:1372
Nuclear power 7:1374
Nuclear weapons 7:1381
Particle accelerators 8:1475
Periodic function 8:1485
Photochemistry 8:1498
Photoelectric effect 8:1502
Physics 8:1513
Pressure 8:1570
Quantum mechanics 8:1607
Radar 8:1613
Radiation 8:1619

Radiation exposure 8:1621
Radio 8:1626
Radioactive tracers 8:1629
Radioactivity 8:1630
Radiology 8:1637
Relativity, theory of 9:1659
Sonar 9:1770
Spectroscopy 9:1792
Spectrum 9:1794
Subatomic particles 10:1829
Superconductor 10:1849
Telegraph 10:1863
Telephone 10:1866
Television 10:1875
Temperature 10:1879
Thermal expansion 5:842
Thermodynamics 10:1885
Time 10:1894
Transformer 10:1908
Transistor 10:1910
Tunneling 10:1937
Ultrasonics 10:1941
Vacuum 10:1960
Vacuum tube 10:1961
Video recording 10:1968
Virtual reality 10:1969
Volume 10:1999
Wave motion 10:2014
X ray 10:2033

Primatology

Animal 1:145
Endangered species 5:793
Mammals 6:1222
Primates 8:1571
Vertebrates 10:1967

Psychiatry/Psychology

Addiction 1:32
Alcoholism 1:85
Attention-deficit hyperactivity disorder (ADHD) 2:237
Behavior 2:270
Cognition 3:511
Depression 4:630
Eating disorders 4:711
Multiple personality disorder 7:1305
Obsession 7:1405
Perception 8:1482
Phobia 8:1497
Psychiatry 8:1592
Psychology 8:1594
Psychosis 8:1596
Reinforcement, positive and negative 9:1657
Savant 9:1712
Schizophrenia 9:1716
Sleep and sleep disorders 9:1745
Stress 9:1826

Radiology

Nuclear medicine 7:1372
Radioactive tracers 8:1629
Radiology 8:1637
Ultrasonics 10:1941
X ray 10:2033

Robotics

Automation 2:242
Mass production 7:1236
Robotics 9:1690

Seismology

Earthquake 4:702
Volcano 10:1992

Sociology

Adaptation 1:26
Aging and death 1:59

Alcoholism **1:**85
Behavior **2:**270
Gerontology **5:**999
Migration (animals) **7:**1271

Technology

Abrasives **1:**2
Adhesives **1:**37
Aerosols **1:**43
Aircraft **1:**74
Alloy **1:**110
Alternative energy sources **1:**111
Antenna **1:**153
Artificial fibers **1:**186
Artificial intelligence **1:**188
Asbestos **1:**191
Automation **2:**242
Automobile **2:**245
Balloon **1:**261
Battery **2:**268
Biotechnology **2:**309
Brewing **2:**352
Bridges **2:**354
CAD/CAM **2:**369
Calculator **2:**370
Canal **2:**376
Cathode **3:**415
Cathode-ray tube **3:**417
Cell, electrochemical **3:**436
Cellular/digital technology **3:**439
Centrifuge **3:**445
Ceramic **3:**447
Compact disc **3:**531
Computer, analog **3:**546
Computer, digital **3:**547
Computer software **3:**549
Cybernetics **3:**605
Dam **4:**611
Diesel engine **4:**646
Diode **4:**665
DVD technology **4:**684
Dyes and pigments **4:**686
Fiber optics **5:**870
Fluorescent light **5:**886
Food preservation **5:**890
Forensic science **5:**898
Generator **5:**962
Glass **5:**1004
Hand tools **6:**1036
Hologram and holography **6:**1048
Incandescent light **6:**1087
Industrial Revolution **6:**1097
Integrated circuit **6:**1106
Internal-combustion engine **6:**1117
Internet **6:**1123
Jet engine **6:**1143
Laser **6:**1166
LED (light-emitting diode) **6:**1176
Lens **6:**1184
Lock **6:**1192
Machines, simple **6:**1203
Magnetic recording/
audiocassette **6:**1209
Mass production **7:**1236
Mass spectrometry **7:**1239
Microwave communication **7:**1268
Paper **8:**1462
Photocopying **8:**1499
Plastics **8:**1532
Polymer **8:**1563
Prosthetics **8:**1579
Radar **8:**1613
Radio **8:**1626
Robotics **9:**1690
Rockets and missiles **9:**1693
Soaps and detergents **9:**1756
Sonar **9:**1770
Space station, international **9:**1788
Steam engine **9:**1817
Submarine **10:**1834
Superconductor **10:**1849
Telegraph **10:**1863
Telephone **10:**1866

Television **10:**1875
Transformer **10:**1908
Transistor **10:**1910
Vacuum tube **10:**1961
Video recording **10:**1968
Virtual reality **10:**1969

Virology

AIDS (acquired immunodeficiency syndrome) **1:**70
Disease **4:**669
Ebola virus **4:**717
Plague **8:**1518
Poliomyelitis **8:**1546
Sexually transmitted diseases **9:**1735
Vaccine **10:**1957
Virus **10:**1974

Weaponry

Ballistics **2:**260
Biological warfare **2:**287
Chemical warfare **3:**457
Forensic science **5:**898
Nuclear weapons **7:**1381
Radar **8:**1613
Rockets and missiles **9:**1693

Wildlife conservation

Biodiversity **2:**281
Biome **2:**293
Biosphere **2:**304
Drift net **4:**680
Ecology **4:**725
Ecosystem **4:**728
Endangered species **5:**793
Forestry **5:**901
Gaia hypothesis **5:**935
Wetlands **10:**2024

Zoology

Amphibians **1:**134
Animal **1:**145
Arachnids **1:**168
Arthropods **1:**183
Behavior **2:**270
Birds **2:**312
Butterflies **2:**364
Canines **2:**382
Cetaceans **3:**448
Cockroaches **3:**505
Coelacanth **3:**508
Coral **3:**566
Crustaceans **3:**590
Dinosaur **4:**658
Echolocation **4:**720
Endangered species **5:**793
Felines **5:**855
Fish **5:**875
Hibernation **6:**1046
Indicator species **6:**1090
Insects **6:**1103
Invertebrates **6:**1133
Kangaroos and wallabies **6:**1153
Mammals **6:**1222
Metamorphosis **7:**1259
Migration (animals) **7:**1271
Mollusks **7:**1288
Plankton **8:**1520
Primates **8:**1571
Reptiles **9:**1670
Snakes **9:**1752
Sponges **9:**1799
Vertebrates **10:**1967

At

Atmosphere, composition and structure

Earth's atmosphere is composed of about 78 percent nitrogen, 21 percent oxygen, and a 1-percent mixture of minor gases dominated by argon. The atmosphere can be divided into vertical layers. Traveling from the surface of the planet upward, the major layers are the troposphere, stratosphere, mesosphere, thermosphere, and exosphere. Although the atmosphere was formed over billions of years, there is growing concern that present-day human activity may be altering the atmosphere to the point that it may affect Earth's climate.

The atmosphere's past

When Earth formed 4.5 million years ago, its atmosphere was probably composed of hydrogen, methane, and ammonia gases—much like the outer planets in our solar system. Some scientists theorize that this original atmosphere may have been lost when the Sun violently emitted material that swept away this gaseous envelope from around Earth.

It is believed that Earth's current atmosphere then began to form when gases were released by early volcanic activity. These gases included water vapor, carbon dioxide, nitrogen, and sulfur or sulfur compounds. The water vapor formed clouds that continually rained on Earth, forming the oceans. Since carbon dioxide dissolves easily in water, the new oceans gradually absorbed most of it. Early plants on the planet then began absorbing sunlight, water, and the remaining carbon dioxide, releasing oxygen as a by-product (a process known as photosynthesis). Over billions

Words to Know

Exosphere: Final layer of the atmosphere, extending from the top of the thermosphere thousands of miles into space.

Ionosphere: A subregion within the thermosphere, extending from about 50 miles (80 kilometers) to more than 150 miles (400 kilometers) above Earth and containing elevated concentrations of charged atoms and molecules (ions).

Mesosphere: The third layer of the atmosphere, extending from the stratosphere to about 50 miles (80 kilometers) above Earth.

Radiation: Energy in the form of waves or particles.

Stratosphere: The second layer of the atmosphere, extending from the tropopause, or top of the troposphere, to about 30 miles (50 kilometers) above Earth.

Thermosphere: The fourth layer of the atmosphere, extending from the top of the mesosphere and extending about 400 miles (640 kilometers) above Earth.

Troposphere: The atmospheric layer closest to ground level, extending up 5 to 10 miles (8 to 16 kilometers) above Earth.

Ultraviolet radiation: Radiation similar to visible light but of shorter wavelength, and thus higher energy.

of years, oxygen increasingly built up in the atmosphere, finally reaching its present-day percentage.

Atmospheric layers

Ninety-nine percent of the total mass of the atmosphere is contained in the first 40 to 50 miles (65 to 80 kilometers) above Earth's surface. The atmosphere can be divided into layers based on atmospheric temperature and pressure. Before 1900, scientists believed that temperatures dropped evenly as elevation increased. Researchers now know, however, that temperatures level off, then increase, beginning at about 8 miles (14 kilometers) above Earth.

The troposphere. The troposphere extends from ground level to a height between 5 and 10 miles (8 and 16 kilometers) above Earth. The troposphere is thickest over the equator and thinnest over the poles. This layer contains 80 percent of the mass of the atmosphere, including all the air we breath and nearly all the water vapor present in the atmosphere. Clouds and all other weather phenomena occur in the troposphere. Temperatures in this layer drop steadily with increasing altitude, about 3.5°F per 1,000 feet (2°C per 305 meters). At Earth's surface, the temperature average is about 63°F (17°C). At the top of the troposphere, an area known as the tropopause, the temperature stops decreasing, having reached as low as −70°F (−57°C).

The stratosphere. The stratosphere extends upward from the troposphere to about 30 miles (50 kilometers) above Earth. Temperatures in the stratosphere stay fairly constant. They begin to rise only near the top of the layer, an area called the stratopause. Here temperatures are nearly as warm as they are on the surface of Earth.

This warming is due to the presence of the ozone layer, or ozonosphere, within the stratosphere, at about 15 miles (24 kilometers) above Earth. Ozone molecules absorb the Sun's ultraviolet radiation (energy in the form of waves or particles) and transform it into heat energy, which heats up the stratosphere and causes the increased temperature. The presence of ozone in the atmosphere is critically important because it prevents ultraviolet light and other harmful radiation from reaching the surface of the planet.

The mesosphere. The mesosphere extends upward from the stratosphere to about 50 miles (80 kilometers) above Earth. Temperatures decrease sharply in this layer, falling from about 20°F (−6°C) at the base to about −130°F (−90°C) at the top.

The thermosphere. The thermosphere extends upward from the mesosphere to about 400 miles (640 kilometers) above Earth. Temperatures rise dramatically in the thermosphere, reaching 2,700°F (1,480°C). Because of these high temperatures, most meteors that enter Earth's atmosphere disintegrate or burn up in this layer.

The thermosphere contains a region known as the ionosphere, which extends from about 50 miles (80 kilometers) to more than 250 miles (400 kilometers) above Earth. The ionosphere contains a high concentration of ions, or electrically charged particles, that help reflect certain radio signals over great distances. High-speed electrons from the Sun are drawn toward the polar regions by Earth's magnetic field. After entering the thermosphere and colliding with air molecules such as oxygen and nitrogen, these particles become luminous, resulting in the colorful aurora borealis (northern lights) in the Northern Hemisphere and the aurora australis (southern lights) in the Southern Hemisphere.

The exosphere. The exosphere is the final layer of the atmosphere, extending from the top of the thermosphere thousands of miles into space. Temperatures in this layer range from about 570°F (300°C) to over 3,000°F (1,650°C). The atmosphere is no longer considered gaseous at this layer because the lack of gravity allows many gas molecules to float off into space.

The atmosphere's future

Future changes to the atmosphere are difficult to predict. There is much concern, however, about the increase of carbon dioxide and the decrease of ozone in the atmosphere. Carbon dioxide is one of the so-called greenhouse gases. These gases absorb some of the solar energy that radiates off Earth, reflecting it back to the surface before it escapes into space. An increase of carbon dioxide levels in the atmosphere, caused by the burning of fossil fuels such as coal and oil, may lead to an increase in Earth's surface temperatures.

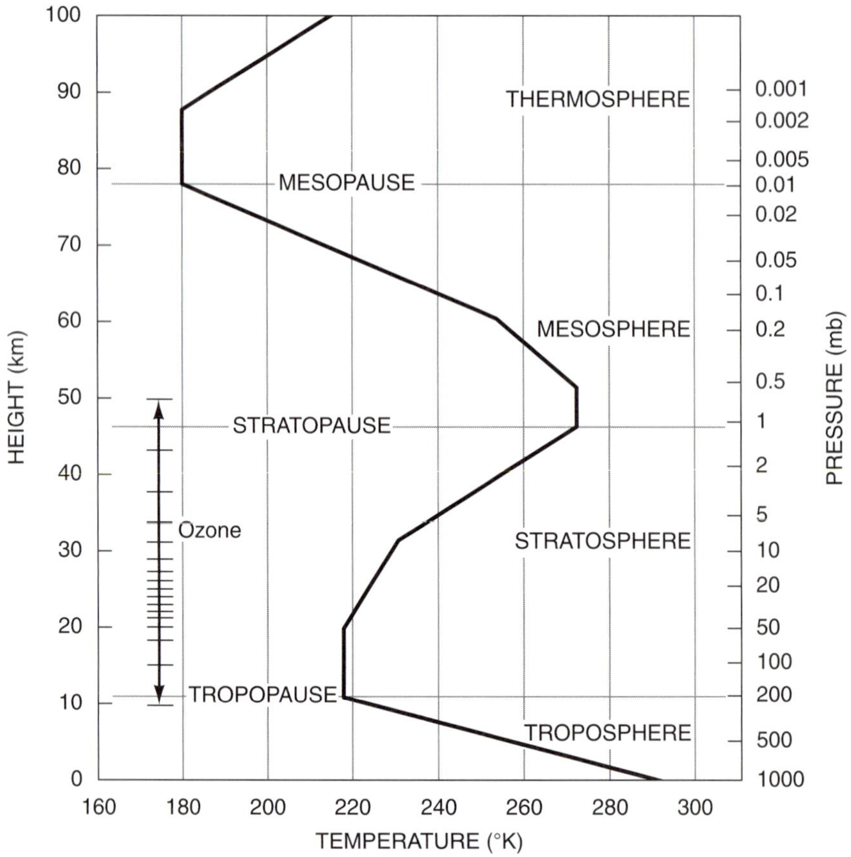

Earth's atmosphere can be divided into layers based on atmospheric temperature and pressure. *(Reproduced by permission of The Gale Group.)*

Decreases in the amount of ozone in the atmosphere were first detected in the mid-1980s. It was discovered afterward that chlorofluorocarbons (CFCs)—industrial chemicals used in refrigerants, aerosol propellants, and solvents—were responsible for destroying ozone. The use of CFCs has since been banned in most industrial countries, but the long-term effects of ozone loss are not yet known.

[*See also* **Atmospheric circulation; Clouds; Greenhouse effect; Hydrologic cycle; Ozone; Weather**]

Atmosphere observation

Atmosphere observation refers to all equipment and techniques used to study properties of the atmosphere, including temperature, pressure, air movements, and chemical composition. Basic instruments for measuring the atmosphere, such as the barometer, were developed during the seventeenth and eighteenth centuries. However, these instruments were useful at first only in studying atmospheric properties close to the ground, not at very high altitudes. Over time, weather instruments were eventually carried into the atmosphere by devices ranging from kites to satellites.

Kites

One of the first means developed for raising instruments to higher altitudes was the kite. In a famous experiment, American statesman Benjamin Franklin used a kite in 1752 to discover that lightning was nothing other than a form of electricity. Within a short period of time, kites were being used by other scientists to carry recording thermometers into the atmosphere, where they could read temperatures at various altitudes.

Weather balloons

An important breakthrough in atmospheric observation came in the late eighteenth century with the invention of the hot-air balloon. Balloon flights made it possible to carry instruments thousands of feet into the atmosphere to take measurements. For the next 150 years, balloons were the primary means by which instruments were lifted into the atmosphere for purposes of observation.

A number of devices were invented for use in weather balloons. The meteorograph was designed to be sent into the atmosphere to automatically record certain measurements, including temperature, pressure, and humidity. The radiosonde is similar in design to the meteorograph, but also includes a radio that can transmit the data collected back to Earth. A radiosonde that is used to collect data about atmospheric winds is known as a rawinsonde.

Balloons are still an important way of transporting weather instruments into the atmosphere. Balloons used to study the properties of the upper atmosphere are known as sounding balloons.

A weather balloon being inflated. *(Reproduced by permission of National Center for Atmospheric Research.)*

Words to Know

Meteorograph: An instrument designed to be sent into the atmosphere to record certain measurements, such as temperature and pressure.

Radiosonde: An instrument for collecting data in the atmosphere and then transmitting that data back to Earth by means of radio waves.

Rawinsonde: A type of radiosonde that is also capable of measuring wind patterns.

Airplanes and rockets

The invention of the airplane and the rocket allowed weather instruments to travel considerably higher than they had ever gone before. Although both can carry the same types of instruments as balloons (only much higher and with greater efficiency), they have taken on tasks more complex than the measurement of atmospheric temperature, pressure, and air movements. For example, airplanes are used to study the properties of hurricanes and to measure levels of ozone and related chemicals over the Antarctic. Such measurements will be valuable in helping scientists to better understand the chemical processes that occur in the atmosphere and the effects they may have on future weather and climate.

Weather satellites

Artificial satellites are the most sophisticated atmospheric observational systems. Lifted into Earth's orbit by a rocket, a weather satellite carries inside it a large number of instruments for measuring many properties of the atmosphere. The first weather satellite ever launched, TIROS 1 (Television and Infrared Observation Satellite), was put into orbit by the U.S. government on April 1, 1960. One of its primary functions was to collect and transmit photographs of Earth's cloud patterns.

Since that time, a number of other weather satellite systems have been put into operation. Satellites can provide a variety of data about atmospheric properties, contributing to improved weather forecasting. As an example, a satellite can track the development, growth, and movement of large storm systems, such as hurricanes and cyclones.

[*See also* **Balloon; Barometer; Global climate; Radar**]

Atmospheric circulation

Atmospheric circulation is the movement of air at all levels of the atmosphere over all parts of the planet. The driving force behind atmospheric circulation is solar energy, which heats the atmosphere with different intensities at the equator, the middle latitudes, and the poles. The rotation of Earth on its axis and the unequal arrangement of land and water masses on the planet also contribute to various features of atmospheric circulation.

Wind cells

There are three wind cells or circulation belts between the equator and each pole: the trade winds (Hadley cells), prevailing westerlies (Ferrell cells), and polar easterlies (polar Hadley cells). The trade winds or Hadley cells are named after the English scientist George Hadley (1685–1768), who first described them in 1753. As air is heated at the equator, it rises in the troposphere, the lowest 10 miles (16 kilometers) of Earth's atmosphere. In the wake of the warm rising air, low pressure develops at the equator. When the air reaches the top of the troposphere, called the tropopause, it can rise no farther and begins to move toward the poles, cooling in the process.

At about 30 degrees latitude north and south, the cooled air descends back to the surface, pushing the air below it toward the equator, since air flows always move toward areas of low pressure. When the north and south trade winds meet at the equator and rise again, an area of calm develops because of the lack of cross-surface winds. Early mariners called this area the doldrums (from an Old English word meaning dull) because they feared their sailing ships would be stranded by the lack of wind.

While most of the trade-wind air that sinks at 30 degrees latitude returns to the equator, some of it flows poleward. At about 60 degrees latitude north and south, this air mass meets much colder polar air (the areas where this occurs are known as polar fronts). The warmer air is forced upward by the colder air to the tropopause, where most of it moves back toward the equator, sinking at about 30 degrees latitude to continue the cycle again. These second circulation belts over the middle latitudes between 30 degrees and 60 degrees are the prevailing westerlies or Ferrell cells, named after the American meteorologist William Ferrell (1817–1891), who discovered them in 1856.

Calm regions also occur at 30 degrees latitude where Hadley cells and Ferrell cells meet because of the lack of lateral wind movement. These

Words to Know

Coriolis effect: Moving object appearing to travel in a curved path over the surface of a spinning body.

Doldrums: Region of the equatorial ocean where winds are light and unpredictable.

Horse latitudes: Region of the oceans around 30 degrees latitude where winds are light and unpredictable.

Jet stream: Rapidly moving band of air in the upper atmosphere.

Polar front: Relatively permanent front formed at the junction of the Ferrell and polar Hadley cells.

Trade winds: Relatively constant wind patterns that blow toward the equator at about 30 degrees latitude.

regions were given the name horse latitudes by sailors bringing horses to the Americas. Stranded by the lack of winds, sailors often ate their horses as supplies ran low.

The air at the top of polar fronts that does not return toward the equator moves, instead, poleward. At the poles, this air cools, sinks, and flows back to 60 degrees latitude north and south. These third circulation belts over the poles are known as polar easterlies or polar Hadley cells because they flow in the same direction as the Hadley cells near the equator. However, they are not as powerful since they lack the solar energy present at the equator.

The Coriolis effect

The air flows in these three circulation belts or cells do not move in a straight north to south or south to north route. Instead, the air flows seem to move east to west or west to east. This effect was first identified by the French mathematician Gaspard-Gustave de Coriolis (1792–1843) in 1835. Coriolis observed that, because of the spinning of the planet, any moving object above Earth's surface tends to drift sideways from its course of motion. In the Northern Hemisphere, this movement is to the right of the course of motion. In the Southern Hemisphere, it is to the left. As a result, surface winds in Hadley cells—both in the equatorial and polar regions—blow from the northeast to the southwest in the Northern Hemisphere and from the southeast to the northwest in the Southern Hemisphere. Surface winds in Ferrell cells tend to blow in the opposite direction: from the southwest to the northeast in the Northern Hemisphere and from the northwest to the southeast in the Southern Hemisphere.

Variations and wind patterns

The conditions of the wind cells described above are for general models. In the real world, actual wind patterns are far more complex. Many elements play a part in disrupting these patterns from their normal course, as described by Hadley and Ferrell. Since the Sun does not always shine directly over the equator, air masses in that area are not heated equally. While some masses in a cell may be heated quickly, creating a strong flow upward, others may not receive as much solar energy, resulting in a much weaker flow. Unevenness in the surface of the planet also affects the movement of air masses in a cell. A mass moving across a uniform region, such as an ocean, may be undisturbed. Once it moves over a region with many variations, such as a mountainous area, it may become highly disturbed.

Doppler radar used to measure the speed and direction of local winds. *(Reproduced by permission of National Oceanic Atmospheric Administration.)*

The jet streams

In 1944, an especially dramatic type of atmospheric air movement was discovered: the jet streams. These permanent air currents are located at altitudes of 30,000 to 45,000 feet (11 to 13 kilometers) and generally move with speeds ranging from about 35 to 75 miles (55 to 120 kilometers) per hour. It is not uncommon, however, for the speed of jet streams to be as high as 200 miles (320 kilometers) per hour.

These narrow tubes of air, which usually travel west to east, are created by the great temperature and pressure differences between air masses. There are four major jet streams, two in each hemisphere. Polar jet streams, formed along the polar front between the Ferrell and polar Hadley cells, move between 30 degrees and 70 degrees latitude. The other jet streams move between 20 degrees and 50 degrees latitude.

Jet streams do not move in straight lines, but in a wavelike manner. They may break apart into two separate streams and then rejoin, or not. In winter, because of greater temperature differences, jet streams are stronger and move toward the equator. In summer, with more uniform temperatures, they weaken and move poleward. The movement of the jet streams is an important factor in determining weather conditions in mid-latitude regions since they can strengthen and move low-pressure systems.

[*See also* **Air masses and fronts; Global climate; Monsoon; Wind**]

Atmospheric optical effects

Rainbows, mirages, auroras, the twinkling of stars, and even the blue color of the sky are all considered atmospheric optical effects. These visual events in the sky occur when light bounces off or is bent by solid particles, liquids droplets, and other materials present in the atmosphere.

Sunlight or white light comprises all the colors of the visible spectrum: red, orange, yellow, green, blue, indigo, and violet. These colors travel at different wavelengths, decreasing in length from red (the longest) to violet (the shortest). When sunlight enters the atmosphere, materials present there break up sunlight into its component colors through reflection (bouncing off an object), refraction (bending through an object), or diffraction (bending around the edge of an object).

Words to Know

Diffraction: The bending of light or another form of electromagnetic radiation as it passes through a tiny hole or around a sharp edge.

Dispersion: The separation of light into its separate colors.

Reflection: The bouncing of light rays in a regular pattern off the surface of an object.

Refraction: The bending of light rays as they pass at an angle from one transparent or clear medium into a second one of different density.

Scattering: The bending of light rays as they bounce off very small objects.

Spectrum: The band of colors that is formed when white light passes through a prism or is broken apart by some other means.

Wavelength: The distance between two troughs or two peaks in any wave.

Rainbows

Rainbows are among the most remarkable effects in the atmosphere. When sunlight enters a raindrop, it is refracted or bent. Since the colors of the spectrum are all bent at different angles, this refraction causes the colors to disperse or separate, as in a prism. The full spectrum of colors is then reflected (bounced) off the back of the raindrop into the air. As the colors pass out of the raindrop into the air, they are refracted a second time. This second refraction causes the different bands of colors to become more distinct. A rainbow is produced as the net result of this sequence of events repeated over and over when the atmosphere is filled with billions of tiny raindrops (such as after a storm).

Mirages

Mirages are one of the most familiar optical effects produced by refraction. Mirages occur when light passes through air layers of different temperatures. One type of mirage—the inferior mirage—forms when a layer of air close to the ground is heated more strongly than the air immediately above it. Light rays passing through the two layers are refracted. As a result, light from the sky appears just below the horizon, like a body of shimmering water. Nearby objects such as trees appear to be reflected in that water.

A second type of mirage—the superior mirage—forms when a layer of air next to the ground is much cooler than the air above it. In this situation, light rays from an object are refracted in such a way that an object on the ground appears to be suspended in air above its true position.

Auroras

Auroras are one of nature's most stunning displays, appearing as luminous streamers, arcs, curtains, or shells in the night sky. Those that occur in the Northern Hemisphere are known as aurora borealis or northern lights; those in the Southern Hemisphere are known as aurora australis or southern lights.

Auroras are produced when electrons (negatively charged particles) from the Sun enter Earth's upper atmosphere. Drawn by Earth's magnetic field to the polar regions, the electrons collide with oxygen and nitrogen molecules present in the atmosphere 50 to 600 miles (80 to 1,000 kilometers) above ground level. As a result of this collision, the molecules become ionized (electrically charged) and glow. Oxygen molecules at

An aurora borealis display. *(Reproduced by permission of Photo Researchers, Inc.)*

lower levels in the atmosphere glow yellow-green. Those at higher levels glow red. Nitrogen molecules glow blue.

Twinkling of stars

As a rule, stars twinkle, but planets do not. Stars are so far away that their light reaches Earth as a single point of light. As that very narrow beam of light passes through Earth's atmosphere, it is refracted and scattered by molecules and larger particles of matter. To an observer on the ground, the star's light appears to twinkle, or blink on and off many times per second.

Sky color

Most of the gas molecules that compose the atmosphere are oxygen and nitrogen. When sunlight strikes these molecules, the longer wavelengths of that light pass right through, but the shorter wavelengths (violet, indigo, blue, and green) are reflected. When we look at the sky, we only see these scattered bands, which combine to appear as varying shades of blue.

Green flashes

In the moment following sunset or sunrise, a flash of green light lasting no more than a second can sometimes be seen on the horizon on the upper part of the Sun. This green flash is the very last remnant of sunlight refracted by Earth's atmosphere, after all red, orange, and yellow rays have disappeared. The green light remains at this moment because the light rays of shorter wavelength—blue and violet—have been scattered by the atmosphere. Green flashes occur very rarely.

Haloes and sun dogs

The passage of sunlight through cirrus clouds can produce optical effects such as haloes and sun dogs. Cirrus clouds, which occur above 16,500 feet (5,000 meters), are made up of millions of ice crystals. Each tiny ice crystal acts like a glass prism, refracting sunlight most commonly at an angle of 22 degrees. Sunlight refracted in this way forms a circle of light—a halo—around the Sun (refracted moonlight forms a halo around the Moon). When relatively large ice crystals are oriented horizontally in a cirrus cloud, the refraction pattern they form is not a circle, but a reflected image of the Sun. Known as sun dogs, these reflected images are located at a distance of 22 degrees from the actual Sun, often at or just above the horizon. Sun dogs are also known as mock suns or parhelia.

Coronas

A corona is a bright disk of light that appears around the Moon or the Sun. It forms when moonlight or sunlight passes through a thin cloud layer filled with water droplets. When the light passes by the edges of the droplets, it is diffracted or bent. This diffraction causes the "white" light to break up into its component colors. Blue light (shortest wavelength) is bent the most, forming the inner ring of the disk. Red light (longest wavelength) forms the outer ring. The disk may be fairly sharp and crisp, or it can be diffuse and hazy. Coronas around the Sun are more difficult to observe because of the Sun's brightness.

[*See also* **Color; Diffraction; Light; Spectrum**]

Atmospheric pressure

Earth's atmosphere consists of gases that surround the surface of the planet. Like any gas, which is made up of molecules that are constantly in motion, the atmosphere exerts a force or pressure on everything within it. This force, divided by the area over which it acts, is the atmospheric pressure. The atmospheric pressure at sea level—considered the mean atmospheric pressure—has an average value of 14.7 pounds per square inch or 29.92 inches of mercury (as measured by a barometer). This means that a one-inch-square column of air stretching from sea level to about 120 miles (200 kilometers) into the atmosphere would weigh 14.7 pounds.

Atmospheric pressure decreases with increasing altitude. The reason for this change with altitude is that atmospheric pressure at any point is a measure of the weight, per unit area, of the atmosphere above that point. Higher altitudes have a lower atmospheric pressure because there is less atmosphere weighing down from above. At an altitude of about 3.1 miles (5 kilometers), the atmospheric pressure is half of its value at sea level.

Atmospheric pressure is closely related to weather. Regions of pressure that are slightly higher or slightly lower than the mean atmospheric pressure develop as air circulates around Earth. The air rushes from regions of high pressure to low pressure, causing winds. The properties of the moving air (cool or warm, dry or humid) will determine the weather for the areas through which it passes. Knowing the location of high and low pressure areas is vital to weather forecasting, which is why they are shown on the weather maps printed in newspapers and shown on television.

[*See also* **Atmosphere, composition and structure; Barometer; Weather; Weather forecasting**]

Atom

An atom is the smallest particle of a element that has all the properties of that element. Imagine that you decide to cut a chunk of aluminum metal into half, over and over again. At some point, you would need very small tools to do the cutting, tools smaller than anything that really exists. However, you would eventually get to the very smallest piece of aluminum that still has all the properties of the original chunk. That smallest piece is an atom of aluminum.

History

One of the questions that ancient Greeks thinkers debated was the structure of matter. Is matter, they asked, continuous or discontinuous? That is, in the aluminum example mentioned above, can a person continue to cut a chunk of aluminum into smaller pieces for ever and ever? Or would the person eventually reach some smallest piece of aluminum that could be divided no further?

Two of the philosophers who argued for the latter opinion were Leucippus (born about 490 B.C.) and his student Democritus (c. 470–c. 380 B.C.). It was Democritus, in fact, who first used the word *atomos* to describe the smallest possible particles of matter. *Atomos* means "indivisible" in Greek.

The particle theory of matter was not developed to any great extent for more than 2,000 years. Then, in 1808, English chemist John Dalton (1766–1844) rephrased the theory in modern terms. Dalton thought of atoms as tiny, indivisible particles, similar to ball bearings or marbles. Dalton's theory of atoms satisfactorily explained what was then known about matter; it was quickly accepted by many other (although not all) chemists.

In the two centuries since Dalton first proposed the modern concept of atoms, that concept has undergone some dramatic changes. We no longer believe that atoms are indivisible particles. We know that they consist of smaller units, known as protons, neutrons, and electrons. These particles are called subatomic particles because they are all smaller than an atom itself. Some subatomic particles are capable of being divided into even smaller units known as quarks.

Modern models of the atom

Scientists think of atoms today in mathematical terms. They use mathematical equations to represent the likelihood of finding electrons in various parts of the atom and to describe the structure of the atomic nucleus, in which protons and neutrons exist.

Most people still find it helpful to think about atoms in physical terms that we can picture in our minds. For most purposes, these pictures are good enough to understand what atoms are like. An atom consists of two parts, a nucleus and a set of one or more electrons spinning around the nucleus.

The nucleus is located at the center of an atom. It consists of one or more protons and, with the exception of the hydrogen atom, one or more neutrons. The number of protons in an atom is given the name atomic number. An atom with one proton in its nucleus has an atomic number of 1, while an atom with sixteen protons in its nucleus has an atomic number of 16. The total number of protons and neutrons in a nucleus is called the atom's mass number. An atom with two protons and two neutrons, for example, has a mass number of 4.

The number of electrons located outside the nucleus of an atom is always the same as the number of protons. An atom with seven protons in its nucleus (no matter how many neutrons) also has seven electrons outside the nucleus. Those electrons travel in paths around the nucleus somewhat similar to the orbits followed by planets around the Sun. Each

A computer-generated model of a neon atom. The nucleus, at center, is too small to be seen at this scale and is represented by the flash of light. Surrounding the nucleus are the atom's electrons. *(Reproduced by permission of Photo Researchers, Inc.)*

of these orbits can hold a certain number of electrons. The first orbit, for example, may hold up to two electrons, but no more. The second orbit may hold up to eight electrons, but no more. The third orbit may hold a maximum of 18 electrons.

These limits determine how the electrons in an atom are distributed. Suppose that the nucleus of an atom contains nine protons. Then the atom also contains nine electrons outside the nucleus. Two of the electrons can be in the first orbit around the nucleus, but the other seven must go to the second orbit.

The term electron orbit is not really correct, even if it does help understand what an electron's path looks like. A better term is electron energy level. The closer an electron is to the nucleus of an atom, the less energy it has; the farther away from the nucleus, the more energy it has.

Physical dimensions

An atom and the particles of which it is composed can be fully described by knowing three properties: mass, electrical charge, and spin.

The mass of protons, neutrons, and electrons is so small that normal units of measurement (such as the gram or centigram) are not used. As an example, the actual mass of a proton is 1.6753×10^{-24} g, or 0.000 000 000 000 000 000 001 675 3 grams. Numbers of this size are so inconvenient to work with that scientists have invented a special unit known as the atomic mass unit (abbreviation: amu) to state the mass of subatomic particles. One atomic mass unit (1 amu) is approximately equal to the mass of a single proton. Using this measure, the mass of a neutron is also about 1 amu, and the mass of an electron, about 0.00055 amu.

The mass of an atom, then, is equal to the total mass of all protons, neutrons, and electrons added together. In the case of the oxygen atom, that mass is expressed as follows:

$$\text{mass of oxygen atom} = \text{mass of 8 protons} + \text{8 neutrons} + \text{8 electrons}$$
$$\text{mass of one oxygen atom} = 8\ \text{amu} + 8\ \text{amu} + (8 \times 0.00055\ \text{amu})$$
$$\text{mass of one oxygen atom} = 16.0044\ \text{amu}$$

The total mass of an atom is called its atomic mass or, less accurately, its atomic weight. As you can see, the mass of an atom depends primarily on the mass of its protons and neutrons and is hardly affected by the mass of its electrons.

The actual mass and size of atoms using ordinary units of measurement are both very small. The mass of one oxygen atom measured in grams is 5.36×10^{-23} g or 0.000 000 000 000 000 000 053 6 grams.

The dimensions of an atom and its nucleus are also amazingly small. The distance across the outside of a typical atom is about 10^{-10} m, or 0.000 000 000 1 meters. In contrast, the distance across a nucleus is about 10^{-15} m, or 0.000 000 000 000 001 meters. In another words, an atom is about 100,000 times larger in size than it its nucleus. To get some idea of this comparison, imagine a pea placed in the center of a large football stadium. If the pea represents the nucleus of an atom, the closest electrons in the atom would be spinning around outside the outermost reaches of the stadium's upper seats.

[*See also* **Atomic mass; Atomic theory; Electron; Element, chemical; Matter, states of; Subatomic particles**]

Atomic mass

The atomic mass of an atom is the mass of that atom compared to some standard, such as the mass of a particular type of carbon atom. The terms atomic mass and atomic weight are often used interchangeably, although, strictly speaking, they do not mean the same thing. Mass is a measure of the total amount of matter in an object. Weight is a measure of the heaviness of an object. In general, the term atomic mass is preferred over atomic weight.

Scientists usually do not refer to the actual mass of an atom in units with which we are familiar (units such as grams and milligrams). The reason is that the numbers needed are so small. The mass of a single atom of oxygen-16, for example, is 2.657×10^{-23} grams, or 0.000 000 000 000 000 000 000 026 57 grams. Working with numbers of this magnitude would be very tedious.

History

Early chemists knew that atoms were very small but had no way of actually finding their mass. They realized, however, that it was possible to express the relative mass of any two atoms. The logic was as follows: suppose we know that one atom of hydrogen combines with one atom of oxygen in a chemical reaction. It is easy enough to find the actual masses of hydrogen and oxygen that combine in such a reaction. Research shows that 8 grams of oxygen combine with 1 gram of hydrogen. It follows, then, that each atom of oxygen has a mass eight times that of a hydrogen atom.

This reasoning led to the first table of atomic masses, published by John Dalton (1766–1844) in 1808. Dalton chose hydrogen to be the

Words to Know

Atomic mass unit (amu): A unit used to express the mass of an atom equal to exactly one-twelfth the mass of a carbon-12 atom.

Isotopes: Two or more forms of an element with the same atomic number (same number of protons in their nuclei), but different atomic masses (different numbers of neutrons in their nuclei).

Mass: Measure of the total amount of matter in an object.

Standard: A basis for comparison; with regard to atomic mass, the atom against which the mass of all other atoms is compared.

Weight: The measure of the heaviness of an object.

standard for his table of atomic masses and gave the hydrogen atom a mass of 1. Of course, he could have chosen any other element and any other value for its atomic mass. But hydrogen was the lightest of the elements and 1 is the easiest number for making comparisons.

One problem with which Dalton had to deal was that he had no way of knowing the ratio in which atoms combine with each other. Since there was no way to solve this problem during Dalton's time, he made the simplest possible assumption: that atoms combine with each other in one-to-one ratios (unless he had evidence for some other ratio).

The table Dalton produced, then, was incorrect for two major reasons. First, he did not know the correct combining ratio of atoms in a chemical reaction. Second, the equipment used at the time to determine mass ratios was not very accurate. Still, his table was an important first step in determining atomic masses. Some of the values that he reported in that first table were: nitrogen: 4.2; carbon: 4.3; oxygen: 5.5; phosphorus: 7.2; and sulfur: 14.4.

Within two decades, great progress had been made in resolving both of the problems that troubled Dalton in his first table of atomic masses. By 1828, Swedish chemist Jöns Jakob Berzelius (1779–1848) had published a list of atomic masses that was remarkably similar to values accepted today. Some of the values published by Berzelius (in comparison to today's values) are: nitrogen: 14.16 (14.01); carbon: 12.25 (12.01); oxygen: 16.00 (16.00); phosphorus: 31.38 (30.97); and sulfur: 32.19 (32.07).

Standards

One of the major changes in determining atomic masses has been the standard used for comparison. The choice of hydrogen made sense to Dalton, but it soon became clear that hydrogen was not the best element to use. After all, atomic masses are calculated by finding out the mass ratio of two elements when they combine with each other. And the one element that combines with more elements than any other is oxygen. So Berzelius and others trying to find the atomic mass of elements switched to oxygen as the standard for their atomic mass tables. Although they agreed on the element, they assigned it different values, ranging from 1 to 100. Before long, however, a value of 16.0000 for oxygen was chosen as the international standard.

By the mid-twentieth century, another problem had become apparent. Scientists had found that the atoms of an element are not all identical with each other. Instead, various isotopes of an element differ slightly in their masses. If O = 16.0000 was the standard, scientists asked, did the 16.0000 stand for all isotopes of oxygen together, or only for one of them?

In order to resolve this question, researchers agreed in 1961 to choose a new standard for atomic masses, the isotope of carbon known as carbon-12. Today, all tables of atomic masses are constructed on this basis, with the mass of any element, isotope, or subatomic particle being compared to the mass of one atom of carbon-12.

Modern atomic mass tables

The atomic mass of an element is seldom a whole number. The reason for this is that most elements consist of two or more isotopes, each of which has its own atomic mass. Copper, for example, has two naturally occurring isotopes: copper-63 and copper-65. These isotopes exist in different abundances. About 69.17 percent of copper is copper-63 and 30.83 percent is copper-65. The atomic mass of the element copper, then, is an average of these two isotopes that takes into account the relative abundance of each: 63.546.

Students sometimes wonder what unit should be attached to the atomic mass of an element. For copper, should the atomic mass be represented as 63.546 g, 63.546 mg, or what? The answer is that atomic mass has no units at all. It is a relative number, showing how many times more massive the atoms of one element are compared to the atoms of the standard (carbon-12).

Still, occasions arise when it would be useful to assign a unit to atomic masses. That procedure is acceptable provided that the same unit is always used for all atomic masses. Scientists have now adopted a unit known as the atomic mass unit for atomic masses. The abbreviation for this unit are the letters amu. One may represent the atomic mass of copper, therefore, either as 63.546 or as 63.546 amu.

[*See also* **Atom; Isotope; Mass spectrometry; Periodic table**]

Atomic theory

An atomic theory is a model developed to explain the properties and behaviors of atoms. As with any scientific theory, an atomic theory is based on scientific evidence available at any given time and serves to suggest future lines of research about atoms.

The concept of an atom can be traced to debates among Greek philosophers that took place around the sixth century B.C. One of the questions that interested these thinkers was the nature of matter. Is matter, they asked, continuous or discontinuous? That is, if you could break apart a piece of chalk as long as you wanted, would you ever reach some ultimate particle beyond which further division was impossible? Or could you keep up that process of division forever? A proponent of the ultimate particle concept was the philosopher Democritus (c. 470–c. 380 B.C.), who named those particles *atomos.* In Greek, *atomos* means "indivisible."

Dalton's theory

The debate over ultimate particles was never resolved. Greek philosophers had no interest in testing their ideas with experiments. They preferred to choose those concepts that were most sound logically. For more than 2,000 years, the Democritus concept of atoms languished as kind of a secondary interest among scientists.

Then, in the first decade of the 1800s, the idea was revived. English chemist John Dalton (1766–1844) proposed the first modern atomic theory. Dalton's theory can be called modern because it contained statements about atoms that could be tested experimentally. Dalton's theory had five major parts. He said:

1. All matter is composed of very small particles called atoms.
2. All atoms of a given element are identical.
3. Atoms cannot be created, destroyed, or subdivided.
4. In chemical reactions, atoms combine with or separate from other atoms.

5. In chemical reactions, atoms combine with each other in simple, whole-number ratios to form combined atoms.

(By the term combined atoms, Dalton meant the particles that we now call molecules.)

Dalton's atomic theory is important not because everything he said was correct. It wasn't. Instead, its value lies in the research ideas it contains. As you read through the list above, you'll see that every idea can be tested by experiment.

Late nineteenth- and early twentieth-century atomic models

As each part of Dalton's theory was tested, new ideas about atoms were discovered. For example, in 1897, English physicist J. J. Thomson (1856–1940) discovered that atoms are not indivisible. When excited by means of an electrical current, atoms break down into two parts. One of those parts is a tiny particle carrying a negative electrical charge, the electron.

To explain what he had discovered, Thomson suggested a new model of the atom, a model widely known as the plum-pudding atom. The name comes from a comparison of the atom with a traditional English plum pudding, in which plums are embedded in pudding, as shown in the accompanying figure of the evolution of atomic theory. In Thomson's atomic model, the "plums" are negatively charged electrons, and the "pudding" is a mass of positive charge.

The nuclear atom. Like the Dalton model before it, Thomson's plum-pudding atom was soon put to the test. It did not survive very long. In the period between 1906 and 1908, English chemist and physicist Ernest Rutherford (1871–1937) studied the effects of bombarding thin gold foil with alpha particles. Alpha particles are helium atoms that have lost their electrons and that, therefore, are positively charged. Rutherford reasoned that the way alpha particles traveled through the gold foil would give him information about the structure of gold atoms in the foil.

Rutherford's experiments provided him with two important pieces of information. First, most of the alpha particles traveled right through the foil without being deflected at all. This result tells us, Rutherford concluded, that atoms consist mostly of empty space. Second, a few of the alpha particles were deflected at very sharp angles. In fact, some reflected completely backwards and were detected next to the gun from which they were first produced. Rutherford was enormously surprised. The result, he said, was something like shooting a cannon ball at a piece of tissue paper and having the ball bounce back at you.

According to Rutherford, the conclusion to be drawn from this result was that the positive charge in an atom must all be packed together in one small region of the atom. He called that region the nucleus of the atom. A sketch of Rutherford's nuclear atom is shown in the figure as well.

The planetary atom. One part of Rutherford's model—the nucleus—has turned out to be correct. However, his placement of electrons created some problems, which he himself recognized. The peculiar difficulty is that electrons cannot remain stationary in an atom, as they appear to be in the figure. If they were stationary, they would be attracted to the nucleus and become part of it. (Remember that electrons are negatively charged and the nucleus is positively charged; opposite charges attract.)

But the electrons could not be spinning around the nucleus either. According to a well-known law of physics, charged particles (like electrons) that travel through space give off energy. Moving electrons would eventually lose energy, lose speed, and fall into the nucleus. Electrons in Rutherford's atom could neither be at rest nor in motion.

The evolution of atomic theory. *(Reproduced by permission of The Gale Group.)*

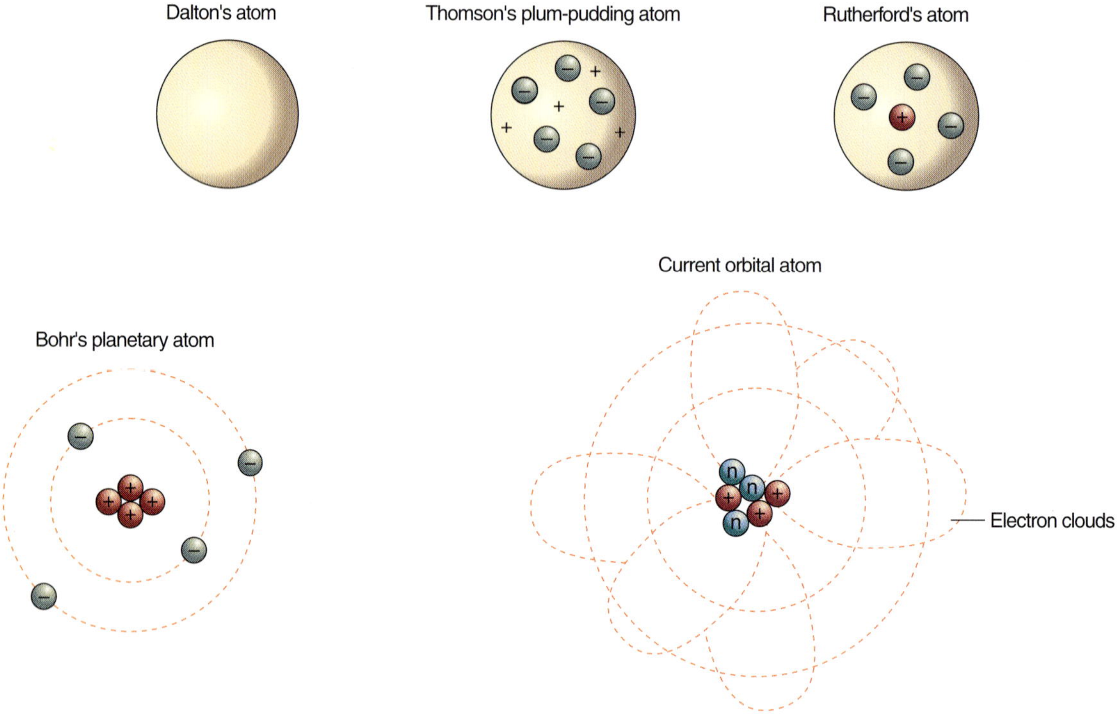

The solution to this dilemma was proposed in a new and brilliant atomic theory in 1913. Suppose, said Danish physicist Niels Bohr (1885–1962), that places exist in the atom where electrons can travel without losing energy. Let's call those places "permitted orbits," something like the orbits that planets travel in their journey around the Sun. A sketch of Bohr's planetary atom is also shown in the figure. If we can accept that idea, Bohr said, the problem with electrons in Rutherford's atom would be solved.

Scientists were flabbergasted. Bohr was saying that the way to explain the structure of an atom was to ignore an accepted principle of physics—at least for certain small parts of the atom. The Bohr model sounded almost like cheating: inventing a model just because it might look right.

The test, of course, was to see if the Bohr model could survive experiments designed specifically to test it. And it did. Within a very short period of time, other scientists were able to report that the Bohr model met all the tests they were able to devise for it. By 1930, then, the accepted model of the atom consisted of two parts, a nucleus whose positive charge was known to be due to tiny particles called protons, and one or more electrons arranged in distinct orbits outside the nucleus.

The neutron. One final problem remained. In the Bohr model, there must be an equal number of protons and electrons. This balance is the only way to be sure that an atom is electrically neutral, which we know to be the case for all atoms. But if one adds up the mass (total amount of matter) of all the protons and electrons in an atom, the total comes no where near the actual mass of an atom.

The solution to this problem was suggested by English physicist James Chadwick (1891–1974) in 1932. The reason for mass differences, Chadwick found, was that the nuclei of atoms contain a particle with no electric charge. He called this particle a neutron.

Chadwick's discovery resulted in a model of the atom that is fairly easy to understand. The core of the atom is the atomic nucleus, in which are found one or more protons and neutrons. Outside the nucleus are electrons traveling in discrete orbits.

Modern theories

This model of the atom can be used to explain many of the ideas in chemistry in which ordinary people are interested. But the model has not been used by chemists themselves for many decades. The reason for this difference is that revolutionary changes occurred in physics during the 1920s. These changes included the rise of relativity, quantum theory, and uncertainty that forced chemists to rethink the most basic concepts about atoms.

As an example, the principle of uncertainty says that it is impossible to describe with perfect accuracy both the position and the motion of an object. In other words, you might be able to say very accurately where an electron is located in an atom, but to do so reduces the accuracy with which you can describe its motion.

By the end of the 1920s, then, chemists had begun to look for new ways to describe the atom that would incorporate the new discoveries in physics. One step in this direction was to rely less on physical models and more on mathematical models. That is, chemists began to give up on the idea of an electron as a tiny particle carrying an electrical charge traveling in a certain direction with a certain speed in a certain part of an atom. Instead, they began to look for mathematical equations which, when solved, gave the correct answers for the charge, mass, speed, spin, and other properties of the electron.

Mathematical models of the atom are often very difficult to understand, but they are enormously useful and successful for professional chemists. The clues they have given about the ultimate structure of matter have led not only to a better understanding of atoms themselves, but also to the development of countless innovative new products in our daily lives.

Do atoms exist?

One of the most remarkable features of atomic theory is that even today, after hundreds of years of research, no one has yet seen a single atom. Some of the very best microscopes have produced images of groups of atoms, but no actual picture of an atom yet exists. How, then, can scientists be so completely certain of the existence of atoms and of the models they have created for them? The answer is that models of the atom, like other scientific models, can be tested by experimentation. Those models that pass the test of experimentation survive, while those that do not are abandoned. The model of atoms that scientists use today has survived and been modified by untold numbers of experiments and will be subjected to other such tests in the future.

[*See also* **Atom; Atomic mass; Electron; Element, chemical; Isotope; Periodic table; Subatomic particle**]

Attention-deficit hyperactivity disorder (ADHD)

Attention-deficit hyperactivity disorder (ADHD) is a condition that is characterized by a person's inability to focus attention. The condition is present at birth and is usually evident by early childhood, although some persons are not diagnosed until adulthood. ADHD is thought to be a disorder of the functioning of the brain that may be caused by hereditary factors or exposure of the developing fetus to harmful substances.

ADHD is estimated to occur in 3 to 5 percent of school-age children in the United States; boys with the disorder outnumber girls who have it. ADHD is a major cause of poor school performance.

Symptoms of ADHD

For purposes of diagnosis, the symptoms of ADHD are divided into two categories: one describes symptoms related to a person's inability to pay attention; the other describes symptoms related to a person's level of hyperactivity and impulsiveness.

Symptoms of inattentiveness include a child's (1) failure to pay attention to detail, (2) tendency to make careless errors in schoolwork, (3) inability to follow instructions or complete tasks with ease, (4) seeming not to listen when spoken to, (5) having apparent difficulty keeping attention on the subject at hand, (6) frequently losing things necessary for schoolwork or play, and (7) being easily distracted by sights or sounds.

Symptoms of hyperactivity include a child's (1) inability to sit still, (2) running around or climbing when expected to remain seated, (3) excessive talking, and (4) difficulty playing or performing activities quietly. Symptoms of impulsive behavior in social situations include (1) blurting out answers before questions are completed, (2) difficulty waiting for one's turn, and (3) interrupting others.

Effect of ADHD on learning

ADHD is not a learning disability, but it often has a serious effect on learning because of a child's inability to pay attention, follow instructions, remember information, or complete a task. Many people who have this disorder are highly intelligent but may do poorly in school because of the regimentation of traditional classroom settings. In addition,

Words to Know

Behavior modification: A type of therapy that uses learning techniques in an attempt to substitute inappropriate behavior with appropriate behavior.

Hyperactivity: A condition of being overly or abnormally active.

Impulsiveness: Spontaneous action without prior thought.

children with ADHD may have problems making friends because of their tendency to take over activities or talk too much, their inability to follow the rules of games or activities, or other inappropriate behavior.

Treatment of ADHD

In order to effectively treat a child with ADHD, the child, his parents, and his teachers must be educated as to the nature of the disorder and how it affects the child's functioning. Treatment usually involves psychological counseling, behavior modification, providing structured settings and controls, and giving the child frequent praise and rewards for completing tasks and controlling behavior.

Treatment with medication is sometimes effective in relieving symptoms of ADHD. The drugs Ritalin and Dexedrine, which are stimulants, have shown remarkable success in temporarily improving a child or an adult's ability to focus in up to 90 percent of cases. These drugs are only effective in the short-term, however. Once the drug leaves the body or is stopped, symptoms of ADHD return. Other drugs, including certain antidepressants, are also sometimes used to control symptoms.

Australia

Of the seven continents, Australia is the flattest, smallest, and, except for Antarctica, the most arid (dry). Including the southeastern island of Tasmania, the island continent encompasses 2,967,877 square miles (7,686,810 square kilometers). Geographically isolated from other landmasses for millions of years, Australia boasts unique animal species, notably the kangaroo, the koala bear, the platypus, and the flightless emu bird. Outside of a few regions (including lush Tasmania), the continent is dry, bleak, and inhospitable.

Origin and topography of Australia

About 95 million years ago, tectonic forces (movements and pressures of Earth's crust) split Australia from Antarctica and the ancient southern supercontinent of Gondwanaland (which comprised present-day Africa, South America, Australia, Antarctica, and India). Geologists

Australia. *(Reproduced by permission of The Gale Group.)*

estimate that Australia is presently drifting northward at a rate of approximately 18 inches (28 centimeters) per year. Millions of years of erosion have worn down the continent's surface features, giving it a relatively flat, uniform appearance. Because of this monotonous desert and semi-desert flatness, broken only by salt lakes, much of Australia is referred to as outback.

A few lush areas exist on the continent. The northern section of the continent experiences tropical temperatures. Cape York Peninsula, jutting up from the northeast section of the continent, is largely covered by rain forest. And an exception to the uniform flatness of the continent is the Great Dividing Range, an entire expanse of mountain ranges stretching along Australia's east coast.

Great Dividing Range

The Great Dividing Range, which extends 1,200 miles (1,931 kilometers), was created during a 125-million-year period beginning 400 million years ago. The Australian Alps form the southern end of the range and contain Australia's highest peak, Mount Kosciusko, at 7316 feet (2,230 meters). The Great Dividing Range is coursed by rivers and streams. Because the range tends to trap moisture from easterly weather fronts originating in the Pacific Ocean, the landscape west of the range is forbidding and the weather hot and dry.

Uluru and the Henbury Craters

Starkly beautiful mountain ranges—the McDonnell and the Musgrave—punctuate the middle of the Australian continent. Between these ranges lies the world's largest sandstone monolith, Uluru (commonly known as Ayers Rock). Uluru is the most sacred site in the country for Australia's aborigines, the native people of the continent. The monolith, two-thirds of which is believed to be below ground surface, is about 2.2 miles (3.5 kilometers) long and 1,143 feet (349 meters) high. The center of the continent also features the Henbury Craters, one of the largest clusters of meteorite craters in the world. The largest of these depressions, formed by the impact of an extraterrestrial rock, is about 591 feet (177 meters) long and 49 feet (15 meters) deep.

Great Barrier Reef

The world's largest coral formation, the Great Barrier Reef, stretches for about 1,250 miles (2,00 kilometers) along Australia's northeast coast.

Several individual reefs compose the Great Barrier Reef, which is home to unusual marine life. In some places, the reef is more than 400 feet (122 meters) thick. It is separated from the Australian continent by a shallow lagoon 10 to 100 miles (16 to 161 kilometers) wide.

Geology of Tasmania

Tasmania separated from mainland Australia only 10,000 years ago, when sea levels rose after the thawing of the last ice age. The island lies 150 miles (240 kilometers) due south of the southeastern tip of Australia, separated by the Bass Strait. Geologically, the island is similar to the Australian continent. However, rainfall is moderate, and the mountains on the island are covered with dense forests.

Natural resources

It is estimated that Australia has 24 billion tons (22 billion metric tons) of coal reserves. Natural gas fields are found throughout the country, supplying most of Australia's domestic needs. There are commercial gas fields in every state and pipelines connecting those fields to major cities. Australia has trillions of tons of estimated natural gas reserves

Uluru (Ayers Rock) in central Australia measures 1,143 feet (349 meters) high. *(Reproduced by permission of JLM Visuals.)*

trapped underground throughout the continent. Australia supplies much of its oil consumption needs domestically, producing about 25 million barrels per year.

Australia has rich deposits of uranium ore, iron ore, nickel, lead, and zinc. In the case of uranium and iron ore, Australia has billions of tons of reserves. Uranium ore is refined for use as fuel for the nuclear power industry. Gold production in Australia peaked early in the twentieth century, but is still quite substantial. The continent is also well known for its precious stones, particularly white and black opals from the south-central region of Australia.

Automation

Automation is the use of computers and robots to automatically control and operate machines or systems to perform work normally done by humans. Although ideas for automating tasks have been in existence since the time of the ancient Greeks, the development of automation came during the Industrial Revolution of the early eighteenth century. Many of the steam-powered devices built by James Watt, Richard Trevithick, Thomas Savery, Thomas Newcomen, and their contemporaries were simple examples of machines capable of taking over the work of humans. Modern automated machines can be subdivided into two large categories: open-loop machines and closed-loop machines.

Open-loop machines

Open-loop machines are devices that are started, go through a cycle, and then stop. A common example is the automatic dishwashing machine. Once dishes are loaded into the machine and a button pushed, the machine goes through a predetermined cycle of operations: pre-rinse, wash, rinse, and dry. Many of the most familiar appliances in homes today (microwave ovens, coffeemakers, CD players) operate on this basis.

Larger, more complex industrial operations also use open-cycle operations. For example, in the production of a car, a single machine may be programmed to place a side panel in place on the car and then weld it in a dozen or more locations. Each of the steps involved in this process—from placing the door properly to each of the different welds—takes place according to instructions programmed into the machine.

Words to Know

Closed-loop machine: Machine that can respond to new instructions during its operation and make consequent changes in that operation.

Feedback mechanism: Ability of a machine to self-correct its operation by using some part of its output as input.

Feedforward mechanism: Ability of a machine to examine the raw materials that come to it and then decide what operations to perform.

Open-loop machine: Machine that performs some type of operation according to a predetermined program and that cannot adjust its own operation.

Closed-loop machines

Closed-loop machines are devices that are capable of responding to new instructions at some point in their operation. The instructions may come from a human operator or from some part of the operation itself. The ability of a machine to self-correct by using some part of its output (for example, measurements) as input (new instructions determined by those measurements) is known as feedback.

One example of a closed-loop operation is the machine used in the manufacture of paper. Paper is formed when a mixture of pulpy fibers and water is emptied onto a conveyer belt. The water drains off, leaving the pulp on the belt. As the pulp dries, paper is formed. The rate at which the pulpy matter is added to the conveyer belt can be automatically controlled by a machine.

A sensing device at the end of the conveyor belt is capable of measuring the thickness of the paper and reporting back to the pouring machine on the condition of the product. If the paper becomes too thick, the sensor can tell the pouring machine to slow the rate at which the pulpy mixture is added to the belt. If the paper becomes too thin, the sensor can tell the machine to increase the rate at which the material is added.

Other types of closed-loop machines contain sensors, but are unable to make necessary adjustments on their own. Instead, sensor readings are sent to human operators who monitor the machine's operation and input any changes needed. Still other closed-loop machines have feedforward mechanisms. Machines of this type examine the raw materials that come to them and then decide what operations to perform. Letter-sorting machines in post offices are of this type. The machine sorts a letter by reading the zip code on the address and then sending the letter to the appropriate subsystem.

The role of computers in automation

Since the 1960s, the nature of automation has undergone dramatic changes as a result of the development of computers. For many years, automated machines were limited by the amount of feedback data they could collect and interpret. Thus, their operation was limited to a relatively small number of alternatives. A modern computer, however, can analyze a vast number of sensory inputs from a system and decide which of many responses it should make.

Artificial intelligence. Present-day computers have made possible the most advanced forms of automation: operations that are designed to replicate human thought processes. The enormous capability of a computer makes it possible for an automated machine to analyze many more options, compare options with each other, consider possible outcomes for

An automated machine is able to perform tasks that could be dangerous or difficult for humans. *(Reproduced by permission of The Stock Market.)*

various options, and perform basic reasoning and problem-solving steps not contained within the machine's programmed memory. At this point, the automated machine can be said to be approaching the types of mental functions normally associated with human beings, that is, to have artificial intelligence.

The human impact of automation

The impact of automation on individuals and societies has been profound. On one level, many otherwise dangerous, unpleasant, or time-consuming tasks are now being performed by machines. The transformation of the communications industry is one example of the way in which automation has made life better for the average person. Today, millions of telephone calls that would once have had to go through human operators are now handled by automatic switching machines.

Automated systems also make it much easier for people to work in nontraditional settings. They may be able to stay home, for example, and do their jobs by communicating with other individuals and machines by means of highly automated communications systems.

However, automation has also had some negative effects on employment. When one machine can do the work of ten workers, most or all of those people will be out of a job. In many cases, those workers will have to be retrained—often learning newer and higher skills—before they can be reemployed.

[*See also* **Artificial intelligence; Robotics**]

Automobile

No invention in modern times has had as much of an impact on human life as the invention of the automobile. It has become an important influence on the history, economy, and social life of much of the world. In fact, the rapid growth of the United States in the twentieth century can be directly related to the automobile.

Automobiles reach into every aspect of society, from the design of our cities to such personal uses as vacation travel, dining, and shopping. Mass-production techniques, first developed for the automobile, have been adapted for use in nearly every industry. Meanwhile, dozens of industries depend, directly or indirectly, on the automobile. These industries include producers of steel and other metals, plastics, rubber, glass, fabrics, petroleum products, and electronic components.

Structure of the automobile

Hundreds of individual parts make up the essential components of the modern automobile. Much like the human body, these parts are arranged into several systems, each with a different function. Each system is necessary for making the automobile run, keeping it safe, and reducing noise and pollution.

The major systems of an automobile are the engine, fuel system, exhaust system, cooling system, lubrication system, electrical system, transmission, and the chassis. The chassis includes the wheels and tires, the brakes, the suspension system, and the body. These systems will be found in every form of motor vehicle and are designed to interact with and support each other.

Engine. The engine—the "heart" of the automobile—operates on internal combustion, meaning the fuel used for its power is burned inside

Karl Friedrich Benz in his first automobile. *(Reproduced courtesy of the Library of Congress.)*

of the engine. This burning occurs inside cylinders, which contain pistons. The pistons are attached, via a connecting rod, to a crankshaft. Gasoline, the most common automobile fuel, is pulled into the cylinder by the vacuum created as the piston moves down through the cylinder. The gasoline is then compressed up into the cylinder by the upward movement of the piston. A spark is introduced through a spark plug placed at the end of the cylinder. The spark causes the gasoline to explode, and the explosion drives the piston down again into the cylinder. This movement, called the power stroke, turns the crankshaft. A final movement of the piston upward again forces the exhaust gases, the byproducts of the fuel's combustion, from the cylinder. These four movements—intake, compression, power, exhaust—are called strokes. The four-stroke engine is the most common type of automobile engine.

Fuel system. Gasoline must be properly mixed with air before it can be introduced into the cylinder. The combination of gasoline and air creates a greater explosion. The fuel pump draws the gasoline from the gas tank mounted at the rear of the car. The gasoline is drawn into a carburetor on some cars, while it is fuel-injected on others. Both devices mix the gasoline with air (approximately 14 parts of air to 1 part of gasoline) and spray this mixture as a fine mist into the cylinders. Other parts of the fuel system include the air cleaner (a filter to ensure that the air mixed into the fuel is free of impurities) and the intake manifold (distributes the fuel mixture to the cylinders).

Exhaust system. After the fuel is burned in the pistons, the gases and heat created must be released from the cylinder to make room for the next intake of fuel. The exhaust system is also responsible for reducing the noise caused by the explosion of the fuel.

Exhaust gases are released from the cylinder through an exhaust valve. The gases gather in an exhaust manifold before eventually being channeled through the exhaust pipe and muffler and finally out the tailpipe and away from the car. The muffler is constructed with a maze of baffles, specially developed walls that absorb energy (in the form of heat, force, and sound) as the exhaust passes through the muffler.

The burning of fuel creates hazardous gases (hydrocarbons, carbon monoxide, and nitrogen oxide) that are extremely harmful to the engine's components and the environment. The emission control system of a car, linked to the exhaust system, functions in two primary ways. First, it reduces the levels of unburned fuel by burning as much of the exhaust as possible. It does this by returning the exhaust to the fuel-air mixture injected into the cylinders. Second, it uses a catalytic converter (fitted before the muffler) to increase the conversion of the harmful gases to less harmful forms.

Cooling system. The cooling system also maintains the engine at a temperature that will allow it to run most efficiently. A liquid-cooled system is most commonly used. The explosion of fuel in the cylinders can produce temperatures as high as 4000°F (2204°C). Liquid-cooling systems use water (mixed with an antifreeze that lowers the freezing point and raises the boiling point of water) guided through a series of jackets attached around the engine. As the water solution circulates through the jackets, it absorbs the heat from the engine. It is then pumped to the radiator at the front of the car, which is constructed of many small pipes and thin metal fins. This design creates a large surface area that draws the heat from the water solution. A fan attached to the radiator uses the wind created by the movement of the car to cool the water solution further. Temperature sensors in the engine control the operation of the cooling system so that the engine remains in its optimal temperature range.

Lubrication. Without the proper lubrication, the heat and friction created by the rapid movements of the engine's parts would quickly cause it to fail. At the bottom of the engine is the crankcase, which holds a supply of oil. A pump, powered by the engine, carries oil from the crankcase and through a series of passages and holes to all the various parts of the engine. As the oil flows through the engine, it forms a thin layer between the moving parts so they do not actually touch. The heated oil drains back into the crankcase, where it cools. The fumes given off by the crankcase are circulated by the PCV (positive crankcase ventilation) valve back to the cylinders, where they are burned off, further reducing the level of pollution given off by the automobile.

Electrical system. Electricity is used for many parts of the car, from the headlights to the radio, but its chief function is to provide the electrical spark needed to ignite the fuel in the cylinders. The electrical system is comprised of a battery, starter motor, alternator, distributor, ignition coil, and ignition switch. The starter motor is necessary for generating the power to carry the engine through its initial movements. Initial voltage is supplied by the battery, which is kept charged by the alternator. The alternator creates electrical current from the movement of the engine, much as windmills and watermills generate current from the movement of air or water.

Turning the key in the ignition switch draws electrical current from the battery. This current, however, is not strong enough to provide spark to the spark plugs. The current is therefore drawn through the ignition coil, which is comprised of the tight primary winding and the looser secondary winding. The introduction of current between these windings creates a powerful magnetic field. Interrupting the current flow, which happens many times a second, causes the magnetic field to collapse. The collapsing of the magnetic field produces a powerful electrical surge. In this way, the 12-volt current from the battery is converted to the 20,000 volts needed to ignite the gasoline.

Because there are two or more cylinders, and therefore as many spark plugs, this powerful current must be distributed—by the distributor—to each spark plug in a carefully controlled sequence. This sequence must be carefully timed so that the cylinders, and the pistons powering the crankshaft, work smoothly together. For this reason, most present-day automobiles utilize an electronic ignition, in which a computer precisely controls the timing and distribution of current to the spark plugs.

Henry Ford in his first automobile.

Transmission. Once the pistons are firing and the crankshaft is spinning, this energy must be converted, or transmitted, to drive the wheels. The crankshaft spins only within a limited range, usually between 1,000 to 6,000 revolutions per minute (rpm). Although the wheels spin at far lower rpms, the range at which they spin is wider (to accommodate the wide range of driving speeds of an automobile). The gears of the transmission accomplish the task of bringing down the fast-spinning input from the crankshaft to the smaller number of rpms needed by the wheels.

There are two types of transmission: manual and automatic. Automobiles generally have at least three gears, plus a reverse gear (many manual transmissions have four or even five gears). With manual transmission, the driver controls the shifting of the gears. In an automatic transmission, gears are engaged automatically. Both types of transmission make use of a clutch, which allows the gears to be engaged and disengaged.

Chassis. The chassis is the framework to which the various parts of the automobile are mounted. The chassis must be strong enough to bear the weight of the car, yet somewhat flexible in order to sustain the shocks and tension caused by turning and road conditions. Attached to the chassis are the wheels and steering assembly, the brakes, the suspension, and the body.

The steering system allows the front wheels to guide the automobile. The steering wheel is attached to the steering column, which in turn is fitted to a gear assembly that allows the circular movement of the steering wheel to be converted to the straight movement of the front wheels. The gear assembly is attached to the front axle by tie rods. The axle is connected to the hubs of the wheels.

Wheels and the tires around them form the automobile's only contact with the road. Tires are generally made of layers of rubber or synthetic rubber around steel fibers that greatly increase the rubber's strength and ability to resist puncture. Proper inflation of the tires improves fuel efficiency and decreases wear on the tires. When applied to the wheels, brakes provide friction that causes the wheels to stop turning.

The suspension system enables the automobile to absorb the bumps and variations in the road surface, keeping the automobile stable. Most cars feature independent front suspension (the two wheels in front are supported separately). In this way, if one wheel hits a bump while the other wheel is in a dip, both wheels will maintain contact with the road. This is especially important because steering the automobile is performed with the front wheels. More and more cars also feature independent rear suspension, improving handling and the smoothness of the ride.

The main components of the suspension system are the springs and the shock absorbers. The springs suspend the automobile above the wheel, absorbing the bumps in the road surface. As the chassis bounces on the springs, the shock absorbers act to dampen, or quiet, the movement of the springs.

The body of a car is usually composed of steel or aluminum, although fiberglass and plastic are also used. While the body forms the passenger compartment, offers storage space, and houses the automobile's systems, it has other important functions as well. In most instances, its solid structure protects passengers from the force of an accident. Other parts of the car, such as the front and hood, are designed to crumple easily, thereby absorbing much of the impact of a crash. A firewall between the engine and the interior of the car protects passengers in case of a fire. Lastly, the body's design helps to reduce the level of wind resistance as the car moves, allowing the driver better handling ability and improving the efficiency of the engine.

[*See also* **Internal combustion engine**]

B

Bacteria

Bacteria are very small organisms, usually consisting of one cell, that lack chlorophyll (a green pigment found in plants that allows for the production of food). Except for viruses, they are the smallest living things on Earth. Many bacteria are so small that a million of them, laid end-to-end, would measure no more than about five centimeters (two inches). The term bacteria is the plural form of the word bacterium, which represents a single organism.

Bacteria are found everywhere, in the air, soil, water, and inside your body and on your skin. They tend to multiply very rapidly under favorable conditions, forming colonies of millions or even billions of organisms within a space as small as a drop of water.

The Dutch merchant and amateur scientist Anton van Leeuwenhoek (1632–1723) was the first person to observe bacteria and other microorganisms. Using single-lens microscopes of his own design, he described bacteria and other microorganisms (calling them "animalcules") in a series of letters to the Royal Society of London between 1674 and 1723.

Today, bacteria are classified in the kingdom Procaryotae. This term refers to the fact that bacteria consist of prokaryotic cells, cells that do not contain a nucleus. (A nucleus is a structure that controls a cell's functions and contains genes. Genes carry the deoxyribonucleic acid [DNA] that determines the characteristics passed on from one generation to the next.) The genetic material of bacteria is contained, instead, within a single, circular chain of DNA.

Characteristics of bacteria

Bacteria are generally classified into three groups based on their shape. They are described as spherical (coccus), rodlike (bacillus), or spiral or

Words to Know

Aerobic bacteria: Bacteria that need oxygen in order to live and grow.

Anaerobic bacteria: Bacteria that do not require oxygen in order to live and grow.

Bacillus: A type of bacterium with a rodlike shape.

Capsule: A thick, jelly-like material that surrounds the surface of some bacteria cells.

Coccus: A type of bacterium with a spherical (round) shape.

Decomposers: Bacteria that break down dead organic matter.

Fimbriae: Short, hairlike projections that may form on the outer surface of a bacterial cell.

Fission: A form of reproduction in which a single cell divides to form two new cells.

Flagella: Whiplike projections on the surface of bacterial cells that make movement possible.

Pasteurization: A process by which bacteria in food are killed by heating the food to a particular temperature for some given period of time.

Pili: Projections that join pairs of bacteria together, making possible the transfer of genetic material between them.

Prokaryote: A cell that has no distinct nucleus.

Spirilla: A type of bacterium with a spiral shape.

Spirochetes: A type of bacterium with a spiral shape.

Toxin: A poisonous chemical.

Vibrio: A type of bacterium with a comma-like shape.

corkscrew (spirochete [pronounced SPY-ruh-keet] or spirilla). Some bacteria also have a shape like that of a comma and are known as vibrio.

As the drawing of the anatomy of a typical bacterium shows, the cytoplasm of all bacteria is enclosed within a cell membrane that is itself surrounded by a rigid cell wall. Bacteria also produce a thick, jelly-like material on the surface of the cell wall. When that material forms a distinct outside layer, it is known as a capsule.

Many rod, spiral, and comma-shaped bacteria have whiplike limbs, known as flagella, attached to the outside of their cells. They use these flagella for movement by waving them back and forth. Other bacteria move simply by wiggling their whole cell back and forth. Some bacteria are unable to move at all.

Two other kinds of projections found on bacterial surfaces include fimbriae and pili. Fimbriae (pronounced FIM-bree-ay) are tiny bristles that allow bacteria to attach themselves to other objects or to surfaces.

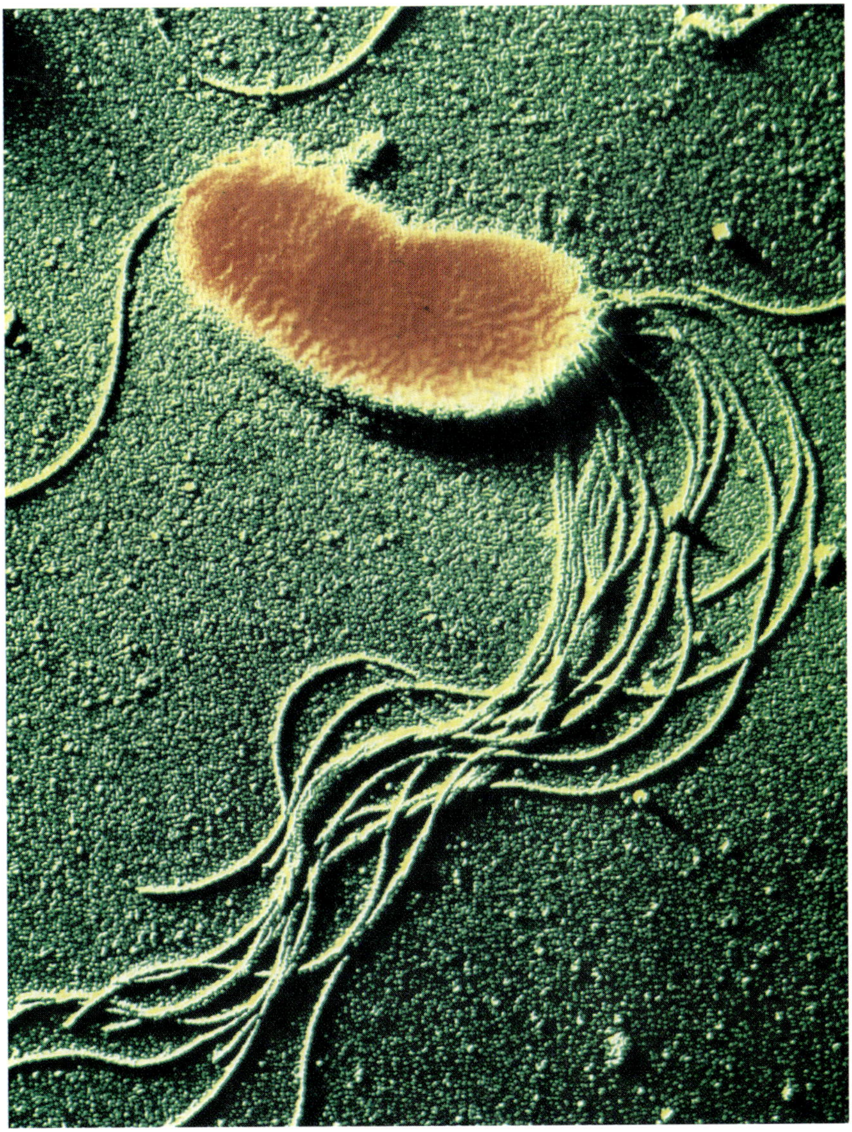

A scanning electron micrograph of the aerobic soil bacterium *Pseudomonas fluorescens*. The bacterium uses its long, whiplike flagellae to propel itself through the water layer that surrounds soil particles. *(Reproduced by permission of Photo Researchers, Inc.)*

Pili are tiny whiskers that allow bacterial cells to exchange genetic material with each other.

Bacterial growth

The term bacterial growth generally refers to the growth of a group of bacteria rather than a single cell. Single cells generally do not get larger in size, so the term growth refers to the reproduction of cells.

Bacteria most commonly reproduce by fission, the process by which a single cell divides to produce two new cells. The process of fission may take anywhere from 15 minutes to 16 hours, depending on the type of bacterium.

A number of factors influence the rate at which bacterial growth occurs, the most important of which are moisture, temperature, and pH. Bac-

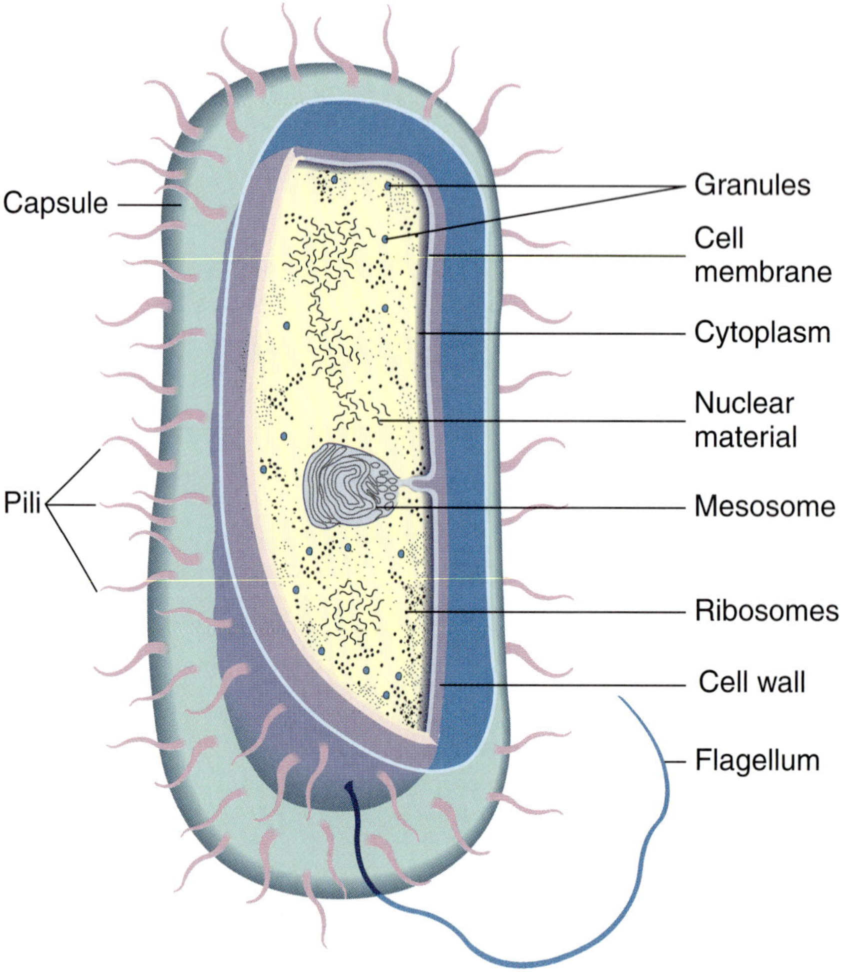

The anatomy of a typical bacterium. *(Reproduced by permission of The Gale Group.)*

teria are about 80 to 90 percent water. If too much water passes into or out of a bacterial cell, the cell dies. The bacterial cell wall provides protection against the gain or loss of water in most ordinary circumstances. But conditions may be such as to produce an unusually large gain or loss of water.

For example, if a bacterial cell is placed in a highly concentrated solution of salt water, water begins to pass out of a cell and into the salt water. The cell begins to shrink and is unable to carry on normal life functions. It cannot grow and will eventually die. On the other hand, an excess of water can be harmful to bacteria also. If water flows into a bacterial cell, the cell begins to swell and may eventually burst, resulting in the death of the cell.

All bacteria have a particular temperature range at which they can survive. For a specific type of bacteria, that range can be very high, very low, or somewhere in between, although it is always a narrow range. Most bacteria thrive at temperatures close to that of the human body (37°C or 98.6°F). But some bacteria prefer cold temperatures as low as freezing (0°C or 32°F), and others require very hot temperatures such as those found in hot springs (50°C to 90°C or 120°F to 200°F). The most extreme conditions in which bacteria have been found are around the hydrothermal vents near the Galapagos Islands. The temperatures near these cracks in the ocean floor is about 350°C (660°F), an environment just right, apparently, for the bacteria that live there.

Another factor affecting bacterial growth is pH, the acidity of a solution. Most bacteria require a pH of 6.7 to 7.5 (slightly more or less acidic than pure water). Other bacteria, however, can survive at a pH more severe than that of battery acid.

Finally, bacteria may or may not require oxygen to grow. Those that do need oxygen are called aerobic bacteria, while those that do not are known as anaerobic bacteria. Anaerobic bacteria have evolved ways of using substances other than oxygen, such as compounds of nitrogen, to obtain the energy they need to survive and grow.

Harmless, beneficial, and harmful bacteria

Bacteria can also be classified according to the effects they have on human life. Some bacteria are used to supply products that improve human life, others cause disease, while still others have no overall affect at all on human life.

Helpful bacteria. Bacteria make possible the digestion of foods in many kinds of animals. Cows, deer, sheep, and other ruminants, for example, have a large organ known as the rumen in which bacteria live and help break down cellulose fibers and other tough plant materials. In humans, bacteria known as *Escherichia coli* (*E. coli*) occur everywhere in the digestive system, aiding in the breakdown of many kinds of foods. Bacteria are also responsible for the production of vitamin K and certain B vitamins.

Certain kinds of bacteria are also essential in the decay and decomposition of waste materials. Such bacteria are known as decomposers. Decomposers attack dead materials and break them down into simpler forms that can be used as nutrients by plants.

Finally, bacteria are involved in the production of many foods eaten humans. For example, bacteria that cause milk to become sour are used in the production of cottage cheese, buttermilk, and yogurt. Vinegar and sauerkraut are also produced by the action of bacteria on ethyl alcohol and cabbage, respectively.

Harmful bacteria. It seems likely, however, that most people know bacteria best because of the diseases they cause. Some of these diseases are produced when bacteria attack directly the tissues in a plant or animal. For example, fruits and vegetables that become discolored as they are growing may be under attack by bacteria.

Bacteria also attack organisms by releasing chemicals that are poisonous to plants and animals. Such poisons are known as toxins. A familiar toxin-producing bacterium is *Clostridium tetani,* responsible for the disease known as tetanus. Tetanus is a condition in which one's muscles are paralyzed, explaining its common name of lockjaw. A related bacterium, *Clostridium botulinum,* releases a toxin that causes the most severe form of food poisoning, botulism.

Some forms of dangerous bacteria live on the human skin, but cause no harm unless they are able to enter the blood stream through a break in the skin. Among these bacteria is *Staphylococcus,* responsible for the potentially fatal toxic shock syndrome. And although *E. coli* is helpful within the digestive system, if it is ingested and enters the bloodstream it causes severe cramping, diarrhea, and possibly even death.

Most forms of food preservation, such as freezing and drying, are designed to kill or inactivate bacteria that would otherwise damage food or cause disease. One of the most common methods of destroying bacteria in foods is pasteurization. Pasteurization is the process of heating a food product to a particular temperature for some given period of time. The temperature and time are selected to be sure that all bacteria in the food are killed by the process. The pasteurization of milk has made it possible to insure safe supplies of one of the most popular of all human foods.

Hardy survivors

In October 2000, a team of biologists claimed to have revived a bacterium that existed 250 million years ago, well before the age of the dinosaurs. They found the bacterium in a drop of fluid trapped in a crystal of rock salt that had been excavated from an air duct supplying a radioactive waste dump 1,850 feet (564 meters) below Earth's surface near Carlsbad, New Mexico. When the biologists drilled into the pocket of fluid in the crystal and mixed nutrients with the fluid, bacteria soon appeared. However, other scientists quickly suggested that the bacteria that grew was simply modern bacteria that had infected the crystal sample. The questioning scientists also pointed out that it would be impossible for the bacterium's DNA (a complex molecule that stores and transmits genetic information) to have survived more than a few thousand years, at best.

Regardless of the debate, bacteria have been around since the dawn of life on Earth, and they have continued to evolve. A major problem facing the medical community today is the ability of disease-causing bacteria to develop a resistance to antibiotics and other antibacterial drugs. These types of bacteria have been able to change their forms or have even been able to secrete enzymes that destroy the antibiotics. Since the

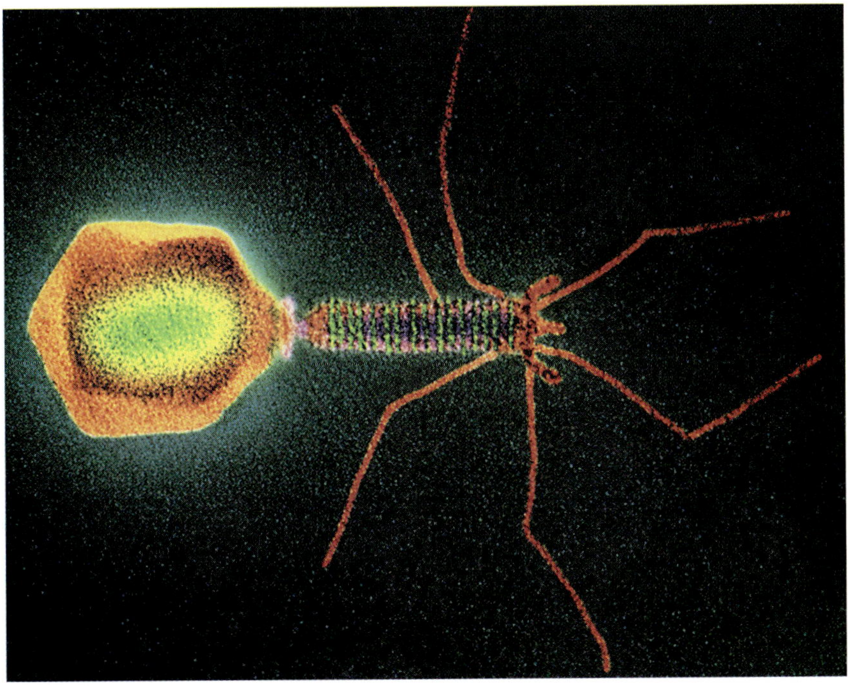

A scanning electron micrograph of a T4 bacteriophage virus. *(Reproduced by permission of Photo Researchers, Inc.)*

development and use of antibiotics in the 1940s, most known bacterial diseases have developed a resistance to at least one type of antibiotic.

[*See also* **Antibiotics; Antiseptics; Fermentation**]

Ballistics

Ballistics is the study of projectile motion. A projectile is an object that has been launched, shot, hurled, thrown, or projected by any other means and that then travels on its own along a ballistic path. For instance, a baseball player throwing a ball from center field to the infield usually throws the ball in a slightly upward direction. The ball's path travels along an arc from the outfield to the infield. Mathematically, the arclike path taken by the ball is known as a parabola.

Ballistics has long been a subject of interest to scientists because bullets, cannon shells, arrows, and other weapons travel in ballistic paths. Military leaders have always valued the information that scientists were able to provide them concerning the proper way in which to aim their guns and bows in combat.

Projectile motion without air resistance

Consider a bullet fired from a rifle that is held parallel to (in the same direction as but never touching) the ground. The path taken by that bullet is affected by two forces. The first force is the velocity given to the bullet by the force of the rifle. (Velocity is the rate at which an object moves in a specified direction; it is measured in meters per second.) That force tends to make the bullet move in a straight line, out of the mouth of the rifle and parallel to the ground. If there were no air present, there would be nothing to slow the motion of the bullet and it would keep traveling with its original velocity.

A second force also operates on the bullet: the force of gravity. As the bullet travels away from the gun, it is pulled downward by Earth's gravitational field. Instead of traveling in a straight line, then, it travels in a curved path towards Earth's surface. That curved path, typical of projectile motion, is a parabola.

The exact shape of the bullet's path is determined by two factors: the mass of the bullet and the velocity with which it travels. The heavier the bullet is, the stronger Earth's gravitational field will pull on it. And the faster the bullet leaves the rifle, the greater its tendency to travel in a straight line away from the gun.

Finding the path for any kind of projectile is an easy problem in physics. If one knows the mass of the object and the velocity with which it is projected, then its pathway can be calculated by well-known formulas.

The practical importance of this calculation is obvious. If a naval ship fires a rocket at an enemy vessel, the path of the rocket must be known. Otherwise the rocket may travel beyond the enemy ship or fall into the water before reaching it. A rocket scientist has to know the path of a space probe launched to Mars if the probe is to land exactly on target rather than sailing on past its intended destination.

Other factors affecting projectile motion

Air resistance is another important factor affecting projectile motion. As a rifle bullet travels through the air, it tends to slow down because of friction between the bullet and the air through which it passes. The amount of friction, in turn, is influenced by a number of factors. Among these factors is the shape of the bullet. Most bullets (and other kinds of projectiles) have pointed front ends—a feature that reduces air resistance. A bullet with a blunt front end would experience a great deal of air resistance and would slow down rapidly. Rotation affects air resistance as well. A good quarterback always tries to place a spin on a football. This helps the football to travel through the air more smoothly than it would without the spin.

Balloon

A balloon is a type of aircraft consisting of a thin envelope filled with a gas less dense than the surrounding air. The envelope can be made of rubber, plastic, treated paper or cloth, or other material through which gases cannot seep. Ordinary party balloons are good models for most kinds of balloon. They are made of rubber that expands when air is blown into them. And the air from a person's breath used to inflate them is less dense than the surrounding air.

Balloon gases

A balloon rises in the air for the same reason that a cork floats in water. Just as a cork's density is less than the density of the surrounding water, so a balloon's density is less than the air around it. The lower-density object, then, is pushed upward by the surrounding higher-density medium.

It stands to reason, then, that the best gas to use in a balloon is the one with the lowest density: hydrogen. In fact, hydrogen was used in the construction of balloons for more than a century. But this gas has one serious drawback. It burns easily and, under the proper circumstances, can even explode. The tragic fire that destroyed the *Hindenburg* dirigible (airship) in 1937 occurred when lightning set fire to hydrogen gas inside it.

Because of hydrogen's flammability, the most popular gas for filling commercial balloons is helium, the second-least dense gas after hydrogen. Helium has 93 percent of the lifting capacity of hydrogen with none of its safety concerns. The problem is that helium is more expensive than is hydrogen. Still, balloons used for commercial and research purposes today almost always use helium as the lifting gas.

Another gas used in balloons is hot air. Hot air has the same chemical composition as ordinary air but, because of its temperature, is less dense that the air around it. Balloons used for sight-seeing and sport usually use hot air. The gondola (traveling compartment) below the balloon itself contains a heating unit that warms air and then pushes it up into the balloon.

Balloon guidance

Vertical (upward or downward) movement of a balloon is generally under human control. In a sight-seeing balloon, the operator can turn the heater on and off to produce more or less hot air. Changes in the amount of hot air make the balloon rise or fall. Vents in the balloon envelope also make it possible to control the amount of gas inside the balloon, therefore changing its vertical movement. The horizontal movement of a balloon is beyond human control, however. Once a balloon has left Earth's surface, its horizontal motion is dependent on wind currents.

History

Joseph (1740–1810) and Jacques Montgolfier (1745–1799) are considered the fathers of ballooning. In 1783, after a series of experiments, the brothers constructed a balloon large enough to carry two humans into the atmosphere, the first manned aircraft.

The suitability of balloons for making atmospheric observations soon became evident, and manned balloon trips soon became common. In 1804, for example, French physicist Joseph Louis Gay-Lussac (1778–1850) traveled to an altitude of 23,000 feet and collected a sample of air. He found that air at that altitude was identical to air at Earth's surface.

In his ascent, Gay-Lussac nearly reached the limits of manned balloon trips without special protection. In contrast, in 1863, two English scientists, James Glaisher and Henry Tracey Coxwell (1819–1900), traveled to a height of over 33,000 feet to study the properties of the upper atmosphere. At that height, the air is so thin that the two men nearly lost their lives.

The greatest of the early balloonists, however, was French meteorologist Léon Philippe Teisserenc de Bort (1855–1913). Over a three-year period between 1899 and 1902, de Bort launched 236 balloons with instruments designed to measure atmospheric conditions.

Scientists are now able to use life-support systems, such as those that are common in space flights, to travel to higher and higher reaches of the atmosphere. The current record is held by two Americans, Ross and Prather, who reached an altitude of 113,740 feet in 1961.

Applications of balloon flight

Balloons are used today primarily for two purposes: for collecting information needed for making weather forecasts and for scientific research. Weather balloons typically carry packages of instruments called radiosondes for measuring the temperature, pressure, density, and other

Hot-air balloon over Albuquerque, New Mexico. *(Reproduced by permission of Phototake.)*

properties of air at some altitude. Each day, thousands of these radiosonde-carrying balloons (called balloonsondes) measure all possible characteristics of the atmosphere around the world. Meteorologists depend on this information for making short- and long-term weather forecasts.

Balloons are also used extensively for astronomical research. Their advantage is that they can take telescopes high enough into the atmosphere that they will not be affected by dust, water vapor, smoke, and other forms of air pollution. Telescopes with a diameter of up to three feet are placed on platforms which are supported by mammoth balloons as tall as eight-story buildings. These telescopes have been carried to altitudes of 120,000 feet.

The Russian mission to Venus in 1985 used two helium balloons to study the motion of the Venusian atmosphere. For 46 hours, they floated above Venus with an attached package of scientific equipment that analyzed the environment and transmitted information directly to Earth.

The success of balloons on Venus has raised the possibility of a similar mission to Mars. American scientists have designed a device consisting of a large hot-air balloon and a much smaller helium-filled bal-

The balloon that carried the 3-ton, 36-inch Stratoscope II telescope into the atmosphere. *(Reproduced by permission of National Aeronautics and Space Administration.)*

loon joined to each other. During the day, the air balloon, heated by the Sun, would drift in the Martian atmosphere with a payload of instruments. At night, the air balloon would cool and descend to the ground, where it would stay, supported in the upright position by the smaller gas balloon. Thus, the same probe would perform the on-ground experiments at night and the atmospheric experiments during the day, traveling from one location to another.

[*See also* **Aerodynamics; Aircraft; Buoyancy**]

Barometer

A barometer is an instrument for measuring atmospheric pressure. Two kinds of barometers are in common use, a mercury barometer and an aneroid barometer. The first makes use of a long narrow glass tube filled with mercury supported in a container of mercury, and the second makes use of an elastic disk whose size changes as a result of air pressure.

Mercury barometers

The principle of the mercury barometer was discovered by the Italian physicist Evangelista Torricelli in about 1643. That principle can be illustrated as follows: a long glass tube is sealed at one end and then filled with liquid mercury metal. The filled tube is then turned upside down and inserted into a bowl of mercury, called a cistern. When this happens, a small amount of mercury runs out of the tube into the cistern, leaving a vacuum at the top of the tube. Vacuums, by nature, exert very little or no pressure on their surrounding environment.

As atmospheric pressure pushes down on the surface of the mercury in the cistern, that mercury in turn pushes up with an equal pressure on the mercury in the glass tube. The height of the mercury in the tube, therefore, reflects the total pressure exerted by the surrounding atmosphere. Under normal circumstances, the column of mercury in the glass tube stands at a height of about 30 inches (76 centimeters) when measured at sea level.

In theory, a barometer could be made of any liquid whatsoever. Mercury is chosen, however, for a number of reasons. It is so dense that the column supported by air pressure is of a usable height. A similar barometer made of water, in comparison, would have to be more than 34 feet (100 meters) high. Mercury also has a low vapor pressure, meaning it does not evaporate very easily. Water has a greater vapor pressure. Because of this, the pressure exerted by water vapor at the top of the

Words to Know

Altimeter: An aneroid barometer used to measure altitude.

Barograph: An aneroid barometer modified to give a continuous reading of atmospheric pressures on graph paper.

Vapor pressure: The amount of pressure exerted by liquid molecules in the vapor state.

barometer would affect the level of the mercury in the tube and the barometric reading, a factor of almost no consequence with a mercury barometer.

Aneroid barometer

A major disadvantage of the mercury barometer is its bulkiness and fragility. The long glass tube can break easily, and mercury levels may be difficult to read under unsteady conditions, as on board a ship at sea. To resolve these difficulties, the French physicist Lucien Vidie invented the aneroid ("without liquid") barometer in 1843.

An aneroid barometer is a container that holds a sealed chamber from which some air has been removed, creating a partial vacuum. An elastic disk covering the chamber is connected to a needle or pointer on the surface of the container by a chain, lever, and springs. As atmospheric pressure increases or decreases, the elastic disk contracts or expands, causing the pointer to move accordingly.

One type of aneroid barometer has a pointer that moves from left to right in a semicircular motion over a dial, reflecting low or high pressure. The simple clocklike aneroid barometer hanging on the wall of many homes operates on this basis. Another type of aneroid barometer has the pointer resting on the side of a rotating cylinder wrapped with graph paper. As the cylinder rotates on its own axis, the pointer makes a tracing on the paper that reflects increases and decreases in pressure. A recording barometer of this design is known as a barograph.

The altimeter. An important application of the aneroid barometer is the altimeter, an instrument used to measure one's distance above sea level. Atmospheric pressure is a function of altitude. The higher one is above sea level, the less the atmospheric pressure; the closer one is to sea level, the greater the atmospheric pressure. A simple aneroid barometer can be used to confirm these differences. If the barometer were mounted in an airplane, a balloon, or some other device that travels up and down in the atmosphere, one could determine the altitude by noting changes in atmospheric pressure.

[*See also* **Atmospheric pressure**]

An aneroid barometer. *(Reproduced by permission of The Stock Market.)*

Battery

A battery is a device for converting chemical energy into electrical energy. Batteries can consist of a single voltaic cell or a series of voltaic cells joined to each other. (In a voltaic cell, electrical energy is produced as the result of a chemical reaction between two different metals immersed in a solution, usually a liquid.) Batteries can be found everywhere in the world around us, from the giant batteries that provide electrical energy in spacecraft to the miniature batteries that power radios and penlights.

The correct use of the term battery is reserved for groups of two or more voltaic cells. The lead storage battery found in automobiles, for example, contains six voltaic cells. However, in common usage, a single cell is often referred to as a battery. For example, the common dry cell battery found in flashlights is really a single voltaic cell.

Types of batteries

Batteries can be classified as primary or secondary batteries (or cells). A primary battery is one designed to be used just once. When the battery has run down (produced all the energy it can), it is discarded. Secondary batteries, on the other hand, can be recharged and reused.

This type of battery is called a carbon-zinc primary cell. A carbon rod sits upright in the center and acts as a cathode. The outer vessel is made of zinc, which allows it to serve as both a container and as an anode. *(Reproduced by permission of Photo Researchers, Inc.)*

Special Kinds of Batteries

Battery	Type	Uses and Special Properties
Zinc/manganese alkaline	Primary	High efficiency: radios, shavers, electronic flash, movie cameras, tape recorders, television sets, clocks, calculators, toys, watches
Mercuric oxide/zinc	Primary	High energy: watches, hearing aids, walkie-talkies, calculators, microphones, cameras
Silver oxide/zinc	Primary	Constant voltage: watches, hearing aids, cameras, calculators
Lithium/copper monofluoride	Primary	High voltage, long shelf life, good low temperature performance: cameras and small appliances
Lithium/sulfur	Primary	Good cold weather performance: emergency power units
Nickel/cadmium	Secondary	Constant voltage and high current: portable hand tools and appliances, shavers, toothbrushes, photoflash equipment, tape recorders, radios, television sets, cassette players and recorders, calculators, pagers, laptop computers
Silver/zinc	Secondary	High power with low weight: underwater equipment, atmospheric and space applications
Sodium/sulfur	Secondary	High temperature performance

The best known example of a primary battery is probably the common dry cell invented by French engineer Georges Leclanché (1839–1882). The dry cell consists of a zinc container that supplies electrons to the battery; a carbon rod through which the electrons flow; and a moist paste of zinc chloride and ammonium chloride, which accepts the electrons produced from the zinc. Technically, the zinc container is the anode (the electrode at which electrons are given up to a reaction) and the moist paste is the cathode (the electrode at which electrons are taken up from a reaction) in the cell. The Leclanché cell is called a dry cell because no liquid is present in it. However, it is not really dry because of the presence of the moist paste, which is needed if electrons are to flow through the cell.

The dry cell runs down as the zinc can is slowly used up. At some point there is not enough zinc left to produce electrons at a useable rate. At that point, the dry cell is just thrown away.

Secondary cells. The secondary cell with which you are likely to be familiar is the lead storage battery found in automobiles. The lead storage battery usually consists of six voltaic cells connected to each other. The total amount of energy produced by the battery is equal to the sum of the electrical energy from the six cells. Since each cell produces about two volts, the total energy available from the cell is 12 volts.

As the lead storage battery is used, it runs down. That is, the lead plates in the battery are converted to lead sulfate. Unlike the dry cell, however, this process can be reversed. Electrical current can be passed back into the battery, and lead sulfate is changed back into lead. If you could see the lead plates in a battery, you would see them slowly disappearing when the battery is being used and slowly reappearing when the battery is being recharged. Recharging occurs naturally when the automobile is operating and generating its own electricity or when a source of external current is provided in order to recharge the battery.

[*See also* **Cell, electrochemical; Electrical conductivity; Electric current; Electricity**]

Behavior

Behavior is the way that all organisms or living things respond to stimuli in their environment. Stimuli include chemicals, heat, light, touch, and gravity. For example, plants respond with growth behavior when light strikes their leaves. Behavior can be categorized as either instinctive (present in a living thing from birth) or learned (resulting from experience). The distinction between the two is often unclear, however, since learned behavior often includes instinctive elements. Plants and animals that lack a well-developed nervous system rely on instinctive behavior. Higher-developed animals use both instinctive and learned behavior. Generally, behavior helps organisms survive.

Plant behavior

The instinctive behavior of a plant depends mainly on growth or movement in a given direction due to changes in its environment. The

Words to Know

Ethology: The scientific study of animal behavior under natural conditions.

Operant conditioning: Trial-and-error learning in which a random behavior is rewarded and subsequently retained.

Stimulus: Something that causes a behavioral response.

Tropism: The growth or movement of a plant toward or away from a stimulus.

growth or movement of a plant toward or away from an external stimulus is known as tropism. Positive tropism is growth toward a stimulus, while negative tropism is growth away from a stimulus. Tropisms are labeled according to the stimulus involved, such as phototropism (light) and gravitropism (gravity). Plants growing toward the direction of light exhibit positive phototropism. Since roots grow downward (with gravity),

Acts of aggression by animals toward one another can be caused by reasons ranging from the protection of their young to territory disputes. *(Reproduced by permission of The Stock Market.)*

they exhibit positive gravitropism. Stems of plants grow upward (against gravity), exhibiting negative gravitropism.

Animal behavior

The scientific study of animal behavior under natural conditions, known as ethology, focuses on both instinctual and learned behavior. Ethologists look at an animal's environment to see how events in that environment combine with an animal's instincts to shape overall behavior. This is especially important in the developing or early stages of an animal's life.

Animals exhibit various levels of instinctual behavior. On a elementary level are reflexes. A reflex is a simple, inborn, automatic response of a part of the body to a stimulus. Reflexes help animals respond quickly to a stimulus, thus protecting them from harm. Other instinctual behaviors are more complex. Examples of this kind include the nest-building behavior of birds and the dam-building behavior of beavers.

Imprinting. An example of animal behavior that combines instinct and learning is imprinting, often seen in birds such as geese and ducks. Within a short, genetically set time frame an animal learns to recognize and then bond to its parent, helping it to survive its infancy. Newly hatched geese or goslings are able to walk at birth. They quickly learn to recognize the movements of their parents and then follow them. If the parents are removed within the first few days after birth and are replaced by

A Canadian goose with her goslings in the Ottawa National Wildlife Refuge in Ohio. *(Reproduced by permission of Field Mark Publications.)*

any moving object, the goslings imprint or bond to that object, learning to follow it.

Animals often add to their set of instinctual behaviors through trial-and-error learning, known as operant conditioning. Young chimps, for example, watch their parents strip a twig and then use the prepared stick to pick up termites from rotten logs. When the young chimps repeat this procedure, their behavior is rewarded by the meal of termites, a preferred food. This reward teaches the chimps to repeat the same behavior when next hungry.

Courtship behaviors. There are many kinds of interactive behavior between animals. One of them is courtship behavior, which enables an animal to find, identify, attract, and arouse a mate. During courtship, animals use rituals, a series of behaviors performed the same way by all the males or females in a species. These include leaping and dancing, singing, the ruffling of feathers, or the puffing up of pouches. The male peacock displays his glorious plumage to the female. Humpback whales announce their presence under the sea by singing a song that can be heard hundreds of miles away.

Group behaviors. Some animals live together in groups and display social behavior. The group protects its members from predators, and allows cooperation and division of labor. Insects, such as bees, ants, and termites, live in complex groups in which some members find food, some defend the colony, and some tend to the offspring. Hierarchies or ranking systems help reduce fighting in a group. Chickens, for example, have a peck-order from the dominant to the most submissive. Each chicken knows its place in the peck-order and does not challenge chickens of higher rank, thereby reducing the chances of fighting. Interactions among group members get more complex with more intelligent species such as apes.

[*See also* **Brain; Nervous system**]

Big bang theory

The big bang is the foremost model that scientists use to describe the creation of the universe. This theory proposes that the universe was created in a violent event approximately 12 to 15 billion years ago. In that event, the lightest elements were formed, which provided the building blocks for all of the matter that exists in the universe today. A consequence of the big bang is that we live in an expanding universe, the ultimate fate of which cannot be predicted from the information we have at this time.

The evolution of the universe

Cosmologists (scientists who study the origin of the universe) believe the universe began as an infinitely dense, hot fireball. They call this single point that contained all the matter in the universe a singularity. Time began at the moment this fireball exploded, stretching space as it expanded rapidly. (Space into which the fireball exploded did not exist separately, but was a part of the fireball at the beginning.) The universe, at first no bigger than the size of a proton, expanded within a microsecond to the size of a basketball. Gravity came into being, and subatomic particles flooded the universe, slamming into one another, forming protons and neutrons (elementary particles that form atoms).

Three minutes after the big bang, the temperature of the universe had cooled to 500,000,000°F (277,777,760°C). Protons and neutrons began to combine to form the nuclei of the simple chemical elements hydrogen, helium, and lithium. Five hundred thousand years later, atoms formed. Some 300 million more years passed before the universe ex-

An artist's impression of galaxies being formed in the aftermath of the big bang. The spiral clouds of gas have already begun condensing into the shapes of future galaxies. *(Reproduced by permission of Photo Researchers, Inc.)*

panded and cooled enough for stars and galaxies to form. Our solar system, formed from a cloud of dust and gas, came into being a mere four-and-a-half billion years ago.

The search for the beginning

A key assumption on which the big bang theory rests is that the universe is expanding. Prior to the twentieth century, astronomers assumed that the universe had always existed as it was, without any changes. In the 1920s, however, American astronomer Edwin Hubble (1889–1953) discovered observable proof that other galaxies existed in the universe besides our Milky Way galaxy. In 1929, he made his most important discovery: all matter in the universe was moving away from all other matter. This proved the universe was expanding.

Hubble reached this conclusion by looking at the light coming toward Earth from distant galaxies. If these galaxies were indeed moving away from Earth and each other, the light they emitted would be stretched or would have a longer wavelength. Since light with a longer wavelength has a reddish tone, this stretching is called redshift. Hubble measured the redshift for numerous galaxies and found not only that galaxies were moving away from Earth in all directions, but that farther galaxies seemed to be moving away at a faster rate.

Inflationary theory and the cosmic microwave background

By the mid-1960s, the big bang theory had received wide acceptance from scientists. However, some problems with the theory still remained. When the big bang occurred, hot radiation (energy in the form of waves or particles) given off by the explosion expanded and cooled with the universe. This radiation, known as the cosmic microwave background, appears as a weak hiss of radio noise coming from all directions in space. It is, in a sense, the oldest light in the universe. When astronomers measured this cosmic microwave background, they found its temperature to be just under −450°F (−270°C). This was the correct temperature if the universe had expanded and cooled since the big bang.

But the radiation seemed smooth, with no temperature fluctuations. If the radiation had cooled at a steady rate, then the universe would have had to expand and cool at a steady rate. If this were true, planets and galaxies would not have been able to form because gravity, which would help them clump together, would have caused fluctuations in the temperature readings.

In 1980, American astronomer Alan Guth proposed a supplemental idea to the big bang theory. Called the inflationary theory, it suggests that at first the universe expanded at a much faster rate than it does now. This concept of accelerated expansion allows for the formation of the stars and planets we see in the universe today.

COBE and MAP

Guth's inflationary theory was supported in April 1992, when NASA (National Aeronautics and Space Administration) announced that its Cosmic Background Explorer (COBE) satellite had discovered those fluctuations. COBE looked about 13 billion light-years into space (hence, 13 billion years into the past) and detected tiny temperature fluctuations in the cosmic microwave background. Scientists regard these fluctuations as proof that gravitational disturbances existed in the early universe, which allowed matter to clump together to form large stellar bodies such as galaxies and planets.

In late 2000, scientists added further supporting evidence to the validity of the big bang theory when they announced that they had analyzed light from a quasar that was absorbed by a distant cloud of gas dust billions of years ago. At that time, the universe would have been about one-sixth of its present age. Based on their findings, the scientists estimated that the background temperature at that point was about −443°F (−264°C), a temperature mark that agrees with the prediction of the big bang theory.

Present-day astronomers liken the study of the cosmic microwave background in cosmology to that of DNA (deoxyribonucleic acid; the complex molecule that stores and transmits genetic information) in biology. They consider it the seed from which stars and galaxies grew. To widen the scope and precision of that study, NASA launched a satellite called the Microwave Anisotropy Probe (MAP) in 2001. Orbiting farther away from Earth than COBE, the goal of MAP is to measure temperature differences in the cosmic microwave background on a much finer scale. Astronomers hope the information gather by MAP will reveal a great deal about the universe, including its large-scale geometry.

[*See also* **Cosmology; Redshift**]

Binary star

A binary star, often called a double star, is a star system in which two stars linked by their mutual gravity orbit around a central point of mass.

Words to Know

Astrometric binary: Binary system in which only one star can be seen, but the wobble of its orbit indicates the existence of another star in orbit around it.

Eclipsing binary: Binary system in which the plane of the binary's orbit is nearly edgewise to our line of sight, so that each star is partially of totally hidden by the other as they revolve around a common point of gravity.

Mass: The quantity of matter in the star as shown by its gravitational pull on another object.

Radiation: Energy in the form of waves or particles.

Spectroscopic binary: A binary system that appears as one star producing two different light spectra.

Spectrum: Range of individual wavelengths of radiation produced when light is broken down by the process of spectroscopy.

Visual binary: Binary system in which each star can be seen directly, either through a telescope or with the naked eye.

Binary stars are quite common. A recent survey of 123 nearby Sun-like stars showed that 57 percent had one or more companions.

English astronomer William Herschel (1738–1822) made the first discovery of a true binary system in the 1700s. He observed the motion of a pair of stars and concluded that they were in orbit around each other. Herschel's discovery provided the first evidence that gravity exists outside our solar system.

Herschel discovered more than 800 double stars. He called these star systems binary stars. His son, John Herschel (1792–1871), continued the search for binaries and catalogued over 10,000 systems of two or more stars.

Types of binary systems

Several kinds of binary stars exist. A visual binary is a pair in which each star can be seen directly, either through a telescope or with the naked eye. In an astrometric binary, only one star can be seen, but the wobble of its orbit indicates the existence of another star in orbit around it. An eclipsing binary is a system in which the plane of the binary's orbit is nearly edgewise to our line of sight. Thus each star is partially or totally hidden by the other as they revolve.

Sometimes a binary system can be detected only by using a spectroscope (a device for breaking light into its component frequencies). If a single star gives two different spectra (range of individual wavelengths of radiation), it is actually a pair of stars called a spectroscopic binary.

A binary star may be a member of one or more of these classes. For example, an eclipsing binary may also be a spectroscopic binary if it is bright enough so that its light spectrum can be photographed.

The only accurate way to determine a star's mass is by studying its gravitational effect on another object. Binary stars have proven invaluable for this purpose. The masses of two stars in a binary system can be determined from the size of their orbit and the length of time it takes them to revolve around each other.

[*See also* **Black hole; Brown dwarf; Doppler effect; Gravity and gravitation; X-ray astronomy**]

An X-ray image of the X-ray binary star system LMC X-1 in the Large Magellanic Cloud (LMC). LMC X-1 is seen as the two bright objects at center left and center right. *(Reproduced by permission of Photo Researchers, Inc.)*

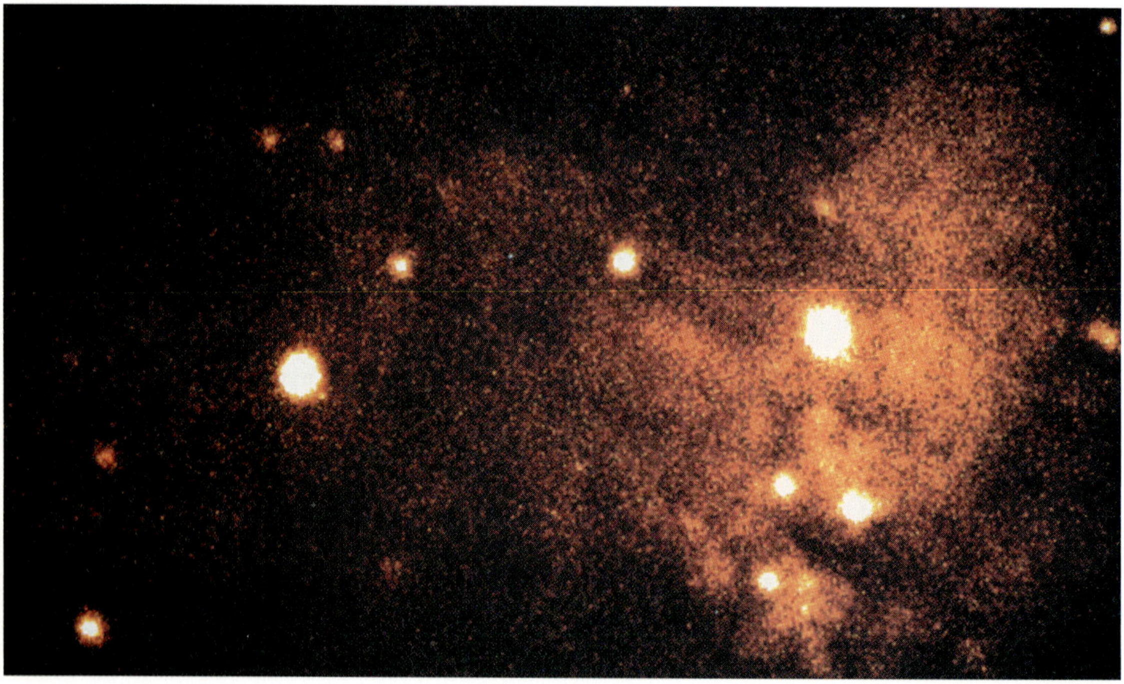

Biochemistry

Biochemistry is the science dealing with the chemical nature of the bodily processes that occur in all living things. It is the study of how plants, animals, and microbes function at the level of molecules.

Biochemists study the structure and properties of chemical compounds in the cells of living organisms and their role in regulating the chemical processes (collectively called metabolism) that are necessary to life. These chemical processes include transforming simple substances from food into more complex compounds for use by the body, or breaking down complex compounds in food to produce energy. For example, amino acids obtained from food combine to form protein molecules, which are used for cell growth and tissue repair. One very important type of protein are enzymes, which cause chemical reactions in the body to proceed at a faster rate.

Complex compounds in food, such as proteins, fats, and carbohydrates, are broken down into smaller molecules in the body to produce energy. Energy that is not needed immediately is stored for later use.

Biochemistry also involves the study of the chemical means by which genes influence heredity. (A gene is a molecule of DNA, or deoxyribonucleic acid, which is found in the nucleus of cells. Genes are responsible for carrying physical characteristics from parents to offspring.)

Words to Know

DNA (deoxyribonucleic acid): A nucleic acid molecule (an organic molecule made of alternating sugar and phosphate groups connected to nitrogen-rich bases) containing genetic information and located in the nucleus of cells.

Gene: A section of a DNA molecule that carries instructions for the formation, functioning, and transmission of specific traits from one generation to another.

Metabolism: The sum of all the chemical processes that take place in the cells of a living organism.

Proteins: Large molecules that are essential to the structure and functioning of all living cells.

A gene can be seen as a sequence of DNA that is coded for a specific protein molecule. These proteins determine specific physical traits (such as hair color, body shape, and height), body chemistry (such as blood type and metabolic functions), and some aspects of behavior and intelligence. Biochemists study the molecular basis of how genes are activated to make specific protein molecules.

[*See also* **Amino acid; Carbohydrate; Chromosome; Enzyme; Hormones; Lipids; Metabolism; Molecular biology; Nucleic acid; Photosynthesis; Protein**]

Biodegradable

The term biodegradable is used to describe substances that are capable of being broken down, or decomposed, by the action of bacteria, fungi, and other microorganisms. Temperature and sunlight may also play a role in the decomposition of biodegradable substances. When substances are not biodegradable, they remain in the environment for a long time, and, if toxic, may pollute the soil and water, causing harm to plants and animals that live in these environments. Humans can also be affected by drinking water or eating crops contaminated by these toxic substances.

Common, everyday substances that are biodegradable include food wastes, tree leaves, and grass clippings. Many communities now encourage people to compost these materials and use them as humus (decayed organic material in soil) for gardening. Because plant and animal materials are biodegradable, this is one way to for towns and cities to reduce solid waste.

The development of detergents in the 1950s and the problems their surfactants caused (wetting agents that allow water to dissolve greasy dirt) raised the issue of the biodegradability of these chemicals. It was found that bacteria in sewage systems degraded some surfactants very slowly. This resulted in the chemicals being released into lakes and streams not fully decomposed and forming suds in the water. Environmental concerns led to the development of new detergents that are more easily biodegradable.

In efforts to control the use of nonbiodegradable materials, governments and industries have taken various measures. For example, the plastic rings that bind six-packs of soda and beer pose a danger to wildlife, who can becoming entangled in them; these rings must now be biodegradable by law in Oregon and Alaska. Italy has banned all nonbiodegradable plastics. Certain manufacturers have responded to the issue by experimenting with biodegradable packaging of food. Many garbage bags and disposable diapers are now being made using degradable plastics, with the goal of reducing litter, pollution, and danger to wildlife.

[*See also* **Composting; Recycling; Waste management**]

Biodiversity

The term biodiversity refers to the wide range of organisms—plants and animals—that exist within any given geographical region. That region may consist of a plot of land no more than a few square meters or yards, a whole continent, or the entire planet. Most commonly, discussions of biodiversity consider all the organisms that interact with each other in an extended geographical region, such as a tropical rain forest or a subtropical desert.

Concerns about biodiversity are relatively new. Only during the last quarter of the twentieth century did scientists begin to appreciate the vast number of organisms found on Earth and the complex ways in which they interact with each other and with their environments. Biologists have now discovered and named about 1.7 million distinct species of plants and animals. As many as 50 million species, however, are thought to exist.

Biodiversity in the tropics is of special interest since the richness of species found there is so great. According to some estimates, 90 percent of all plant, animal, and insect species exist in tropical regions. At the same time, surveys of organisms in the tropics have been very limited. Those studies that have been conducted provide only a hint of the range of life that may exist there. As an example, one study of a 108-square kilometer (42-square mile) reserve of dry forest in Costa Rica found about 700 plant species, 400 vertebrate species, and 13,000 species of insects. Included among the latter group were 3,140 species of moths and butterflies alone.

Human threats to biodiversity

One reason for the growing interest in biodiversity is the threat that human activities may pose for plant and animal species. As humans take over more land for agriculture, cities, highways, and other uses, natural habitats are seriously disrupted. Whole populations may be destroyed, upsetting the balance of nature that exists in an area. The loss of a single plant, for example, may result in the loss of animals that depend on that plant for food. The loss of those animals may, in turn, result in the loss of predators who prey on those animals.

As human populations grow, the threat to biodiversity will continue to grow with it. And as more people place greater stress on the natural environment, greater will be the loss of resources plant and animal communities need to survive.

Why is biodiversity important?

Maintaining biodiversity in a region and across the planet is important for a number of reasons. First, some people argue that all species—

The Amazonian rain forest is rich in plant and animal life. *(Reproduced by permission of Photo Researchers, Inc.)*

because they exist—have a right to continue to exist in their own natural habitats, untouched by human development. Second, humans depend on many of the plants and animals that make up an ecological community. For example, one-quarter of all the prescription drugs in the United States contain ingredients obtained from plants. And third, humans themselves benefit from the interaction among organisms in a biologically diverse community: plants help clean the water and air, provide oxygen in the atmosphere, and control erosion. "Biodiversity," according to the biologist Peter Raven, "keeps the planet habitable and ecosystems functional."

Protection of threatened biodiversity

One of the great issues in environmental science today is how biodiversity can be preserved both in specific geographical regions and across the planet. One proposal that has been made involves the use of ecological reserves. Ecological reserves are protected areas established for the preservation of habitats of endangered species, threatened ecological communities, or representative examples of widespread communities. By the end of the 1990s, there were about 7,000 protected areas globally with an area of 651 million hectares (1.6 billion acres). Of this total, about 2,400 sites comprising 379 million hectares (936 million acres) were fully protected and could be considered to be true ecological reserves.

Ideally, the design of a national system of ecological reserves would provide for the longer-term protection of all native species and their natural communities including terrestrial (land-dwelling), freshwater, and marine (saltwater) systems. So far, however, no country has put in place a comprehensive system of ecological reserves to fully protect its natural biodiversity. Moreover, in many cases existing reserves are relatively small and are threatened by environmental change, illegal poaching of animals and plants, and tourism.

The World Conservation Union, World Resources Institute, and United Nations Environment Program are three important agencies whose purpose is to conserve and protect the world's biodiversity. These agencies have developed the Global Biodiversity Strategy, an international program to help protect plant and animal habitats for this and future generations. Because this program began only in the late 1970s, it is too early to evaluate its success. However, the existence of this comprehensive international effort is encouraging, as is the participation of most of Earth's countries, representing all stages of economic development.

[*See also* **Ecosystem; Endangered species**]

Bioenergy

Bioenergy or "biomass energy" is any type of fuel or power that is made from living matter or biomass. Biomass is a scientific term for organic or living matter that is available on a renewable basis, such as plants. Today, fuel produced from biomass like agricultural products, forest products, and waste provides more than 3 percent of this nation's energy demands. Bioenergy is clean, renewable energy.

Solar energy supports life

Bioenergy has been described as solar energy stored up in plant matter, and in many ways all of the energy used by living things is ultimately derived from the Sun. The Sun is constantly bombarding Earth with its energy, and about one-tenth of 1 percent of the energy that reaches our surface is "fixed" or captured by green plants using the process of photosynthesis (by which a plant converts light energy into chemical energy or food). This "fixing" of solar energy is the basis of and supports all the life found in Earth's major ecosystems. It provides, by way of the food chain, all of mankind's food either directly or indirectly. So when we eat fruit and vegetables, we are consuming the energy from the Sun that had been captured and converted by a plant. When we eat meat like the beef of a cow, we are at least one step removed from the green plant and are getting the energy from the cow (who got the energy from the plant).

The energy generated by the burning wood in this stove puts forth heat used to cook food and heat the home. *(Reproduced by permission of Corbis-Bettmann.)*

Fossil fuels

Ever since humans learned how to control fire, they have been burning wood to cook and to keep warm—that is, they have been using a form of biomass energy or bioenergy. In our recent history, mankind has been burning "fossil fuels," like coal, oil, and natural gas, instead of wood. These are called fossil fuels because they were formed over a span of millions of years out of the remains of dead animals and plants. Coal, oil, and natural gas can therefore be described as biomass energy in concentrated form. However, since these fossil fuels take millions of years to form, they cannot be considered a renewable resource. In other words, we may run out of then at some point in the future.

Words to Know

Biogas: Methane produced by the decomposition of organic material by bacteria.

Biomass: Any biological material used to produce energy.

Digestor: Sealed, enclosed facility in which bacteria digest or decompose organic material.

Ethanol: An alcohol made by fermenting a biomass that is high in carbohydrates, such as corn or sugarcane.

Fossil fuel: A fuel such as coal, oil, or natural gas that is formed over millions of years from the remains of plants and animals.

Greenhouse effect: The warming of Earth's atmosphere due to water vapor, carbon dioxide, and other gases in the atmosphere that trap heat radiated form Earth's surface.

Besides someday running out, fossil fuels also pose another problem. Burning coal, natural gas, and oil creates serious pollution. When these fuels are burned (in order to get them to release their energy), they release carbon dioxide and other gases into the atmosphere, some of which are known as "greenhouse gases." These gases keep some of the Sun's heat from radiating back off Earth's surface and trap it in much the same way that the glass in a greenhouse does. This natural phenomenon is known as the greenhouse effect. Therefore, the more fossil fuels we burn, the more greenhouse gases we produce, and the more heat we trap—making Earth warmer than it should be. Many believe that this global warming will do more harm than good by causing too much or too little rain, or even by melting the polar ice. Another major effect of burning fossil fuels is the release of methane gas. When dissolved in rain, it makes what is called "acid rain," which is quite harmful to trees as well as to the fish in our lakes, rivers, and streams.

Bioenergy as an alternative

Since fossil fuels present us with major pollution problems (and may eventually run out anyway), and nuclear energy confronts us with the dilemma of what to do with nuclear waste, bioenergy has become an increasingly attractive alternative. Probably the most attractive aspect of bioenergy is that it is truly a renewable energy source. The production of vegetative matter throughout the world continues on its own at an astounding rate and can be supplemented whenever needed by additional planting. Compare the millions of years necessary for the development of fossil fuels to that of a single growing season for grass. Besides being an easily renewable resource, bioenergy is also non-polluting. Where fossil fuels spew large amounts of carbon dioxide into the air, using renewable biomass adds no carbon dioxide. Even better, biomass energy recycles carbon dioxide since plants take it in during photosynthesis and use it to make their own food. Biomass energy does have some disadvantages, such as collecting biomass takes a lot of time, large storage areas are needed, and simply burning it wastes a lot of its heat energy.

Bioenergy fuels

Despite these limitations, bioenergy is becoming increasingly attractive mainly because it can be converted into either liquid or gas forms. Starting with almost any types of fast-growing trees and grasses—as well as leftovers from agriculture or forestry, and the organic parts of city and industrial wastes—biomass can be converted directly into liquid fuels for our transportation needs. The two most common liquid "biofuels" are ethanol and biodiesel. Ethanol is an alcohol made by fermenting a biomass that is high in carbohydrates, such as corn or sugarcane. Fermentation happens when yeast digests sugar and produces alcohol as a by-product (this is how beer and wine are made). The alcohol produced is added to gasoline and is often called gasohol. Biodiesel is made adding vegetable oils or even animals fats like recycled cooking grease to diesel fuel.

Gas can also be produced from biomass and is used to generate electricity. Biogas can be made from materials like sewage and waste that are allowed to be decomposed by bacteria. Under the proper conditions in a sealed facility called a digester, the bacteria go to work and decompose this organic material, thereby producing a flammable gas called methane. This sometimes happens naturally in a landfill, and methane is produced by decaying biomass. The digester, however, captures the methane, then pipes it to a storage tank to be used to run turbines that make electricity. Finally, crops that have a high oil content, like coconuts, sunflowers, and soybeans, can be converted chemically into a fuel oil that can be burned like petroleum to make electricity.

The future of bioenergy

Since the United States imports about half of its oil from parts of the world that are sometimes politically unstable, its Department of Energy (DOE) began a program to step up the investigation of alternative fuels like bioenergy. The DOE sponsors a program of biomass research that involves universities, private companies, and government laboratories across the nation, and as a consequence, U.S. production of ethanol approached 1.5 billion gallons (5.7 billion liters) per year by the end of 2000. The DOE also estimates that if two-thirds of the nation's unused cropland were used to grow what are called energy crops, those 35 million acres (14 million hectares) could produce between 15 and 35 billion gallons (57 and 132 billion liters) of ethanol each year to fuel cars, trucks, and buses. As a natural source of alternative energy, bioenergy is a reliable, safe, and clean fuel, and chances are that in the near future the liquid coming from the gasoline nozzle into your car will have its roots in both a farm field as well as an oil field.

[*See also* **Alternative energy sources**]

Biological warfare

Biological warfare (previously called germ warfare) is the use of disease-causing microorganisms as military weapons. One of the earliest recorded uses of biological weapons occurred in the fourteenth century. Invading Asian armies used a device called a catapult to hurl bodies of plague (a deadly, highly contagious disease caused by a bacterium) victims over city walls to infect the resisting townspeople. It is thought that this practice resulted in the spread of the Black Death throughout Europe, killing millions of people in four years.

Toward the end of the French and Indian Wars in North America (1689–1763), a British military officer is said to have given blankets infected with smallpox germs to a tribe of Native Americans, resulting in their infection with the often fatal disease.

In more modern times, an outbreak of inhalation anthrax (a disease caused by inhaling the spores of the anthrax bacterium) in a city in Russia resulted in over 1,000 deaths in 1979. It is thought that this outbreak may have resulted from an accident at a biological warfare facility.

Biological warfare is among the least commonly used military strategies. Most military leaders have been reluctant to release microorganisms that might cause an uncontrolled outbreak of disease, affecting not only the enemy but friendly populations as well.

Words to Know

Microorganism: An organism so small that it can be seen only with the aid of a microscope.

Plague: A contagious disease that spreads rapidly through a population and results in a high rate of death.

Toxin: A poisonous substance produced by an organism.

Microorganisms used as biological weapons

The microorganisms generally considered suitable for biological warfare include viruses, bacteria, protozoa, and fungi. Toxins (poisonous chemicals) produced by microorganisms also are considered biological weapons. These agents are capable of causing sickness or death in humans or animals, destroying crops, or contaminating water supplies.

Various bacteria have been used or experimented with as biological weapons. Anthrax is an infectious disease that can be passed from cattle and sheep to humans. Inhaling anthrax spores can result in a deadly form of pneumonia. During World War II (1939–45), Japan and Great Britain built and tested biological weapons carrying anthrax spores, and the inhalation of anthrax may still be a threat as a biological weapon today.

The toxin that causes botulism (pronounced BOTCH-uh-liz-um), a rare but deadly form of food poisoning, is regarded as one of the most powerful nerve poisons known to science. Ingestion of a very tiny amount can cause death. The toxin has been tested by the U.S. Army as a coating for bullets and as an ingredient in aerosols (for release into the air).

Brucellosis (pronounced broos-uh-LOHS-us) is a bacterial disease transmitted from animals to humans either by direct contact or by drinking the milk of infected goats and cows. It can be used as a biological weapon that does not kill people but makes them so ill that they are unable to resist an attack.

Saxitoxin is a powerful poison produced by one-celled organisms called dinoflagellates (pronounced dye-no-FLAJ-uh-lets) that live in coastal waters. When present in large numbers, the organisms turn the water a reddish color (called red tides). Shellfish contaminated with saxitoxin can cause partial paralysis or even death in humans who eat them and have been considered for use as biological weapons by American military scientists.

Staphylococcus (pronounced staff-luh-KOCK-us) is any of several strains of bacteria that can cause mild to severe infection in humans. The more dangerous strains are the ones most often tested as possible biological weapons. *Staphylococcus* toxin can be dried and stored for up to a year without losing its effectiveness.

Tularemia (pronounced two-luh-REE-mee-uh) is a plaguelike bacterial disease often transmitted through insect bites. In humans, tularemia can cause fever, chills, headache, chest pain, and difficulty breathing. At one time, the U.S. Army considered tularemia as of the most promising of all biological weapons.

Genetically engineered weapons

The development of genetic engineering in the second half of the twentieth century has presented the possibility of creating even more dangerous forms of existing microorganisms—forms that could be used as biological weapons. Genetic engineering is the process of altering the genetic material of living cells in order to make them capable of (1) manufacturing new substances, (2) performing new functions, (3) being more easily produced, or (4) holding up well under storage.

The use and control of biological weapons

In 1925, the Geneva Protocol, a treaty banning the first use of biological and chemical weapons in war, was signed and ratified or officially approved by many nations, but not Japan or the United States (the U.S. government did not ratify the treaty until April 1975, some 50 years later). The treaty did not, however, prohibit the use of these weapons in response to an initial biological or chemical attack from an opponent.

Following the signing of the treaty, some nations, including Japan and the United States, conducted their own research on biological weapons, explaining that such studies were necessary in order to develop defensive measures against the use of such weapons by others.

The most serious violator of the Geneva Protocol was Japan, The Japanese military used biological warfare during the 1930s and 1940s in its conquest of China. In addition, captured American soldiers were used by Japan during World War II as test subjects in biological weapons experiments.

In the decades following World War II, the United States maintained a large and aggressive program of biological weapons research.

Experiments and tests of biological agents were conducted at dozens of American army bases. In 1969, President Richard Nixon announced that the United States was discontinuing further research on biological and chemical weapons.

In 1972, eighty-seven nations (including the United States) signed the Biological Weapons Convention Treaty, which banned the development, testing, and storage of such weapons. The treaty was entered into force three years later. By 2000, the twenty-fifth anniversary of the treaty being entered into force, over 160 nations had signed the treaty; more than 140 of those had also ratified it. However, in 1982, President Ronald Reagan had declared that the world situation justified research on biological and chemical weapons and that the United States would return to a more ambitious program in this area.

As of the end of the twentieth century, over 450 repositories that sold and shipped plague, anthrax, typhoid fever, and other toxic organisms were located throughout the world. The shipment of pathogenic (disease-causing) organisms is not governed by international laws or treaties. Harmful bacteria can be grown in a garage or a backyard, and guidelines on growing bacteria are instantly available on the Internet. In a typical year, the Federal Bureau of Investigation (FBI) investigates about four times as many biological attack threats in the United States as either chemical or nuclear threats. Clearly, law enforcement by local, state, and federal authorities must be strengthened against the possibility of biological disaster.

[*See also* **Bacteria; Chemical warfare; Poisons and toxins; Virus**]

Biology

Biology (from the Greek *bios,* meaning "life") is the scientific study of all forms of life, including plants, animals, and microorganisms.

Biology is composed of many fields, including microbiology, the study of microscopic organisms such as viruses and bacteria; cytology, the study of cells; embryology, the study of development; genetics, the study of heredity; biochemistry, the study of the chemical structures in living things; morphology, the study of the anatomy of plants and animals; taxonomy, the identification, naming, and classification of organisms; and physiology, the study of how organic systems function and respond to stimulation. Biology often interacts with other sciences, such as psychology. For example, animal behaviorists would need to understand

Words to Know

Classification: The system of arranging plants and animals in groups according to their similarities.

Genetic engineering: Altering hereditary material (by a scientist in a lab) by interfering in the natural genetic process.

Germ theory of disease: The belief that disease is caused by germs.

Microorganism: An organism that cannot be seen without magnification under a microscope.

Molecular biology: A branch of biology that deals with the physical and chemical structure of living things on the molecular level.

Natural selection: Process by which those organisms best adapted to their environment survive and pass their traits to offspring.

the biological nature of the animal they are studying in order to evaluate a particular animal's behavior.

History of biological science

The history of biology begins with the careful observation of the external aspects of organisms and continues with investigations into the functions and interrelationships of living things.

The fourth-century B.C. Greek philosopher Aristotle is credited with establishing the importance of observation and analysis as the basic approach for scientific investigation. He also organized the basic principles of dividing and subdividing plants and animals, known as classification. By A.D. 200, studies in biology were centered in the Arab world. Most of the investigations during this period were made in medicine and agriculture. Arab scientists continued this activity throughout the Middle Ages (400–1450).

Scientific investigations gained momentum during the Renaissance (a period of rebirth of art, literature, and science in Europe from the fourteenth to the seventeenth century). Italian Renaissance artists Leonardo da Vinci (1452–1519) and Michelangelo (1475–1564) produced detailed anatomical drawings of human beings. At the same time, others were dissecting cadavers (dead bodies) and describing internal anatomy. By the seventeenth century, formal experimentation was introduced into the study of biology. William Harvey (1578–1657), an English physician, demonstrated the circulation of the blood and so initiated the biological discipline of physiology.

So much work was being done in biological science during this period that academies of science and scientific journals were formed, the first being the Academy of the Lynx in Rome in 1603. The first scientific journals were established in 1665 in France and Great Britain.

The invention of the microscope in the seventeenth century opened the way for biologists to investigate living organisms at the cellular level—and ultimately at the molecular level. The first drawings of magnified life were made by Francesco Stelluti, an Italian who published drawings in 1625 of a honeybee magnified to 10 times its normal size.

During the eighteenth century, Swedish botanist Carolus Linnaeus (1707–1778) developed a system for naming and classifying plants and animals that replaced the one established by Aristotle (and is still used today). Based on his observations of the characteristics of organisms, Linnaeus created a ranked system in which living things were grouped according to their similarities, with each succeeding level possessing a larger number of shared traits. He named these levels class, order, genus, and species. Linnaeus also popularized binomial nomenclature, giving each living thing a Latin name consisting of two parts—its genus and species—which distinguished it from all other organisms. For example, the wolf received the scientific name *Canis lupus,* while humans became *Homo sapiens.*

In the nineteenth century, many explorers contributed to biological science by collecting plant and animal specimens from around the world. In 1859, English naturalist Charles Darwin (1809–1882) published *The Origin of Species by Means of Natural Selection,* in which he outlined his theory of evolution. Darwin asserted that living organisms that best fit their environment are more likely to survive and pass their characteristics on to their offspring. His theory of evolution through natural selection was eventually accepted by most of the scientific community.

French microbiologist and chemist Louis Pasteur (1822–1895) showed that living things do not arise spontaneously. He conducted experiments confirming that microorganisms cause disease, identified several disease-causing bacteria, and also developed the first vaccines. By the end of the nineteenth century, the germ theory of disease was established by German physician Robert Koch (1843–1910), and by the early twentieth century, chemotherapy (the use of chemical agents to treat or control disease) was introduced. The use of antibiotics became widespread with the development of sulfa drugs in the mid-1930s and penicillin in the early 1940s.

From the nineteenth century until the end of the twentieth century, the amount of research and discovery in biology has been tremendous. Two fields of rapid growth in biological science today are molecular biology and genetic engineering.

[*See also* **Biochemistry; Botany; Ecology; Evolution; Genetics; Molecular biology; Physiology**]

Biome

A biome is an ecosystem containing plant and animal species that are characteristic to a specific geographic region. (An ecosystem is the community of plants and animals in an area considered together with their environment.) The nature of a biome is determined primarily by climate, including a region's annual average temperature and amount of rainfall. Biomes are often named for the vegetation found within them. They can be classified as terrestrial (land), aquatic (water), or anthropogenic (dominated by humans). Some familiar examples of biomes include tundra, desert, chaparral, and open ocean. The accompanying drawing shows the variety of biomes that can be found along just two lines of longitude on Earth's surface.

A number of attempts have been made to classify the world's biomes. One of the best known was proposed by the Russian-born German climatologist Wladimir Koeppen (1846–1940). Some of the biomes described by Koeppen are described below.

Terrestrial biomes

Tundra. A tundra is a treeless region in a cold climate with a short growing season. Most tundras receive little precipitation. Still, their soil may be moist or wet because little evaporation occurs. Loss of water by seepage is also prevented because the soil is frozen. Very little vegetation grows and very few animals live in the coldest, most northern, high-arctic tundras. These tundras are dominated by long-lived but short-statured plants, typically less than 5 to 10 centimeters (2 to 4 inches) tall. Low-arctic tundras are dominated by shrubs as tall as 1 meter (3 feet), while on wet sites relatively productive meadows of sedge, cottongrass, and grass grow. In North America, arctic tundras can support small numbers of plant-eating mammals, such as caribou and musk oxen, and even smaller numbers of their predators, such as wolves.

Words to Know

Anthropogenic: Resulting from the influence of human action on nature.

Aquatic: Related to water.

Benthic: Referring to the deepest parts of the oceans.

Boreal: Located in a northern region.

Conifer: Plants whose seeds are stored in cones and that retain their leaves all year around.

Deciduous: Plants that lose their leaves at some season of the year, and then grow them back at another season.

Ecosystem: An ecological community, including plants, animals, and microorganisms, considered together with their environment.

Eutrophic: A productive aquatic region with a large nutrient supply.

Herbaceous: A type of plant that has little or no woody tissue and usually lives for only one growing season.

Lentic ecosystem: An ecosystem that contains standing water.

Lotic ecosystem: An ecosystem that consists of running water.

Monoculture: An ecosystem dominated by a single species.

Oligotrophic: An unproductive aquatic region with a relatively modest nutrient supply.

Pelagic: Referring to the open oceans.

Polyculture: An ecosystem that consists of a wide variety of species.

Temperate: Mild or moderate.

Tropical: Characteristic of a region or climate that is frost free with temperatures high enough to support—with adequate precipitation—plant growth year round.

Upwelling: The process by which lower, nutrient-rich waters rise upward to the ocean's surface.

Wetlands: Areas that are wet or covered with water for at least part of the year.

Boreal coniferous forest. The boreal coniferous forest, or taiga, is an extensive northern biome occurring in moist climates with cold winters. The boreal forest is dominated by coniferous (cone-bearing) trees, especially species of fir, larch, pine, and spruce. Some broad-leaved trees are also present in the boreal forest, especially species of aspen, birch, poplar, and willow. Most boreal forests are subject to periodic catastrophic disturbances, such as wildfires and attacks by insects.

Temperate deciduous forest. Temperate deciduous forests are dominated by a large variety of broad-leaved trees in relatively moist, temperate (mild or moderate) climates. Because these forests occur in places where the winters can be cold, the foliage of most species is seasonally deciduous, meaning that trees shed their leaves each autumn and then regrow them in the springtime. Common trees of the temperate deciduous forest biome in North America are ash, basswood, birch, cherry, chestnut, dogwood, elm, hickory, magnolia, maple, oak, tulip-tree, and walnut.

Temperate rain forest. Temperate rain forests are characterized by mild winters and an abundance of rain. These systems are too moist to support wildfires. As a result, they often develop into old-growth forests, dominated by coniferous trees of mixed age and various species. Individual trees can be very large and, in extreme cases, can be more than 1,000 years old. Common trees of this biome are species of Douglas-fir, hemlock, cedar, redwood, spruce, and yellow cypress. In North America, temperate rain forests are most commonly found on the humid west coast.

A boreal forest in north Saskatchewan. *(Reproduced by permission of JLM Visuals.)*

Temperate grassland. Temperate grasslands occur under climatic conditions that are between those that produce forests and those that produce deserts. In temperate zones, grasslands typically occur in regions where rainfall is 25 to 60 centimeters (10 to 24 inches) per year. Grasslands in North America are called prairies and in Eurasia they are often called steppes. This biome occupies vast regions of the interior of these continents.

The prairie is often divided into three types according to height of the dominant vegetation: tall grass, mixed grass, and short grass. The once-extensive tall grass prairie is dominated by various species of grasses and broad-leaved, herbaceous plants such as sunflowers and blazing stars, some as tall as 3 to 4 meters (10 to 13 feet). Fire played a key role in preventing much of the tall grass prairie from developing into open forest. The tall grass prairie is now an endangered natural ecosystem because it has been almost entirely converted for agricultural use.

The mixed grass prairie occurs where rainfall is less plentiful, and it supports shorter species of grasses and other herbaceous plants. The short grass prairie develops when there is even less precipitation, and it is subject to unpredictable years of severe drought.

Tropical grassland and savanna. Tropical grasslands are present in regions with as much as 120 centimeters (47 inches) of rainfall per year, but under highly seasonal conditions with a pronounced dry season. Savannas are dominated by grasses and other herbaceous plants. However, they also have scattered shrubs and tree-sized woody plants that form a very open canopy (a layer of spreading branches).

Tropical grasslands and savannas can support a great seasonal abundance of large, migratory animals as well as substantial populations of resident animals. This is especially true of Africa, where on the savanna range—among other animals—gazelles and other antelopes, rhinos, elephants, hippopotamuses, and buffalo, and various predators of these, such as lions, cheetahs, wild dogs, and hyenas.

Chaparral. Chaparral is a temperate biome that develops in environments where precipitation varies widely from season to season. A common chaparral pattern involves winter rains and summer drought, the so-called Mediterranean climate. Chaparral is characterized by dwarf forests, shrubs, and herbaceous vegetation. This biome is highly prone to wildfire. In North America, chaparral is best developed in parts of the southwest, especially coastal southern California.

Opposite Page: Biomes along 87 degrees west longitude and along 0 degrees longitude. *(Reproduced by permission of The Gale Group.)*

Desert. Deserts occur in either temperate or tropical climates. They commonly are found in the centers of continents and in rain shadows of

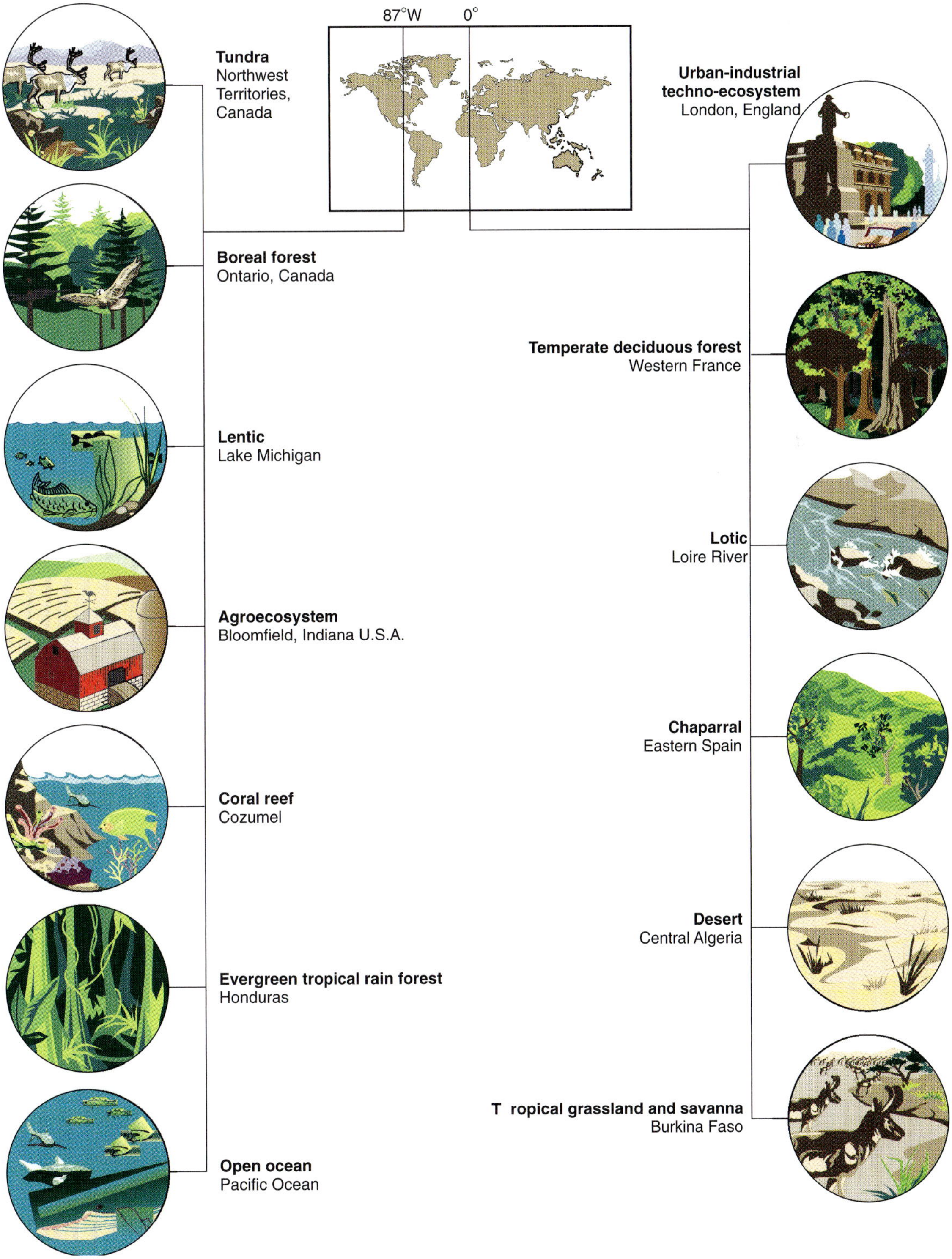
87°W
0°
Tundra
Northwest Territories, Canada
Boreal forest
Ontario, Canada
Lentic
Lake Michigan
Agroecosystem
Bloomfield, Indiana U.S.A.
Coral reef
Cozumel
Evergreen tropical rain forest
Honduras
Open ocean
Pacific Ocean
Urban-industrial techno-ecosystem
London, England
Temperate deciduous forest
Western France
Lotic
Loire River
Chaparral
Eastern Spain
Desert
Central Algeria
T ropical grassland and savanna
Burkina Faso

mountains (a dry region on the side of a mountain sheltered from rain). The most prominent characteristic of a desert is the limited amount of water available. In most cases, less than 25 centimeters (10 inches) of rain fall each year. Not surprisingly, the plant life found in a desert ecosystem is strongly influenced by the availability of water: the driest deserts support almost no plant life, while less-dry deserts may support communities of herbaceous, succulent (having fleshy tissues that conserve water), and annual (returning year after year) plants. In somewhat moister places, a shrub-dominated ecosystem is able to develop.

Semi-evergreen tropical forest. A semi-evergreen tropical forest is a type of tropical forest that develops when a region experiences both wet and dry seasons during the year. Because of this pattern, most trees and shrubs of this biome are seasonally deciduous, meaning that they shed their foliage in anticipation of the drier season. This biome supports a great richness of species of plants and animals, though somewhat less than in tropical rain forests.

Evergreen tropical rain forest. Evergreen tropical rain forests occur in tropical climates with abundant precipitation and no seasonal drought. Because wildfires and other types of catastrophic disturbances are uncommon in this sort of climate regime, tropical rain forests usually develop into old-growth forests. As such, they contain a great richness of species of trees and other plants, a great size range of trees, and an extraordinary diversity of animals and microorganisms. Many ecologists consider the old-growth tropical rain forests the ideal ecosystem on land because of the enormous variety of species that are supported under relatively favorable climatic conditions.

Freshwater biomes

Freshwater biomes can be divided into three general categories: lentic, lotic, and wetlands.

Lentic. A lentic ecosystem is one such as a lake or pond that contains standing water. In lentic systems, water generally flows into and out of the lake or pond on a regular basis. The rates at which inflow and outflow occur vary greatly and can range from days, in the case of small pools, to centuries, in the case of the largest lakes.

The types of organisms that inhabit lentic biomes are strongly influenced by water properties, especially nutrient concentration and water transparency and depth. Waters with a large nutrient supply are highly productive, or eutrophic, while infertile waters are unproductive, or oligotrophic. Commonly, shallow bodies of water are much more productive than deeper bodies of water of the same surface area, primarily because plant growth is influenced by the ability of light to penetrate into the water. Water that becomes cloudy because of the accumulation of silt or dissolved organic matter is likely to have low productivity.

Lotic. A lotic biome is one that consists of running water, as in streams or rivers. The organisms found in a lotic biome depend on factors such as the amount of water in the system, the rate at which it flows, and seasonal changes in the flow rate. Consider a stream in which flooding is common in the spring. Rapidly moving water churns up clay, silt, sand, and other materials from the streambed. The water then becomes cloudy and murky, and light is thus prevented from penetrating it. In this case the stream will not be able to support many kinds of life-forms.

In general, the common lotic ecosystems such as rivers, streams, and brooks are not usually self-supporting in terms of the organisms that live within them. Instead, they typically rely on organic matter carried into them from the land around them or from upstream lakes to support fish and other organisms that live in the biome.

Wetlands. Wetlands are areas that are wet or covered with water for at least part of the year. Some examples of wetlands are marshes, swamps, bogs, sloughs, and fens. Marshes are the most productive wetlands, and are typically dominated by relatively tall plants such as reeds, cat-tails, and bulrushes and by floating-leaved plants such as water lilies and lotus. Swamps are forested wetlands that are seasonally or permanently flooded. In North America, swamps are dominated by tree-sized plants such as bald cypress or silver maple.

Bogs are wetlands that develop in relatively cool but wet climates. They tend to be acidic and, therefore, biologically unproductive. Bogs depend on nutrients obtained from the atmosphere, and are typically dominated by species of sphagnum moss. Fens also develop in cool and wet climates, but they have a better nutrient supply than bogs. Consequently, they are less acidic and more productive than bogs.

Marine biomes

Open ocean. The character of the open-water, or pelagic, oceanic biome is determined by factors such as waves, tides, currents, salinity (salt content), temperature, amount of light, and nutrient concentration. The number of organisms supported by these factors is small and can be compared to some of the least productive terrestrial biomes, such as deserts. The lowest level of food webs in the ocean are occupied by tiny organisms known as phytoplankton. Various species of phytoplankton range in size from extremely small bacteria to larger algae that consist of a single cell and may or may not live in large colonies.

The phytoplankton are grazed upon by small crustaceans known as zooplankton. Zooplankton, in turn, are eaten by small fish. At the top of the pelagic food web are very large predators such as bluefin tuna, sharks, squid, and whales.

The deepest levels of the ocean make up the benthic biome. Organisms in this biome are supported by a meager rain of dead organisms from its surface waters. The benthic ecosystems are not well known, but they appear to be extremely stable, rich in species, and low in nutrient productivity.

Continental shelf waters. Continental shelf waters are areas of ocean water that lie relatively near a coastline. Compared with the open ocean, waters over continental shelves are relatively warm and are well supplied with nutrients from rivers flowing into them. A secondary source of nutrients is water brought to the surface from deeper, more fertile waters that were stirred up by turbulence caused by storms.

Because of the nutrients found in the continental shelf waters, phytoplankton here are relatively abundant and support the larger animals present in the open ocean. Some of the world's most important commercial fisheries are on the continental shelves, including the North and Barents Seas of western Europe, the Grand Banks and other shallow waters of northeastern North America, the Gulf of Mexico, and inshore waters of much of western North America.

Upwelling regions. In certain regions of the ocean, conditions make possible upwellings to the surface of relatively deep, nutrient rich waters. Because of the increased nutrient supply, upwelling areas are relatively fertile, and they support sizeable populations of animals, including large species of fishes and sharks, marine mammals, and seabirds. Some of Earth's most productive fisheries occur in upwelling areas, such as those off the west coast of Peru and other parts of South America and large regions of the Antarctic Ocean.

Estuaries. An estuary is a region along a coastline where a river empties into the ocean. Estuaries display characteristics of both marine and freshwater biomes. They typically have substantial inflows of freshwater from the nearby land, along with large fluctuations of saltwater resulting from tidal cycles. Examples of estuaries include coastal bays, sounds, river mouths, salt marshes, and tropical mangrove forests.

Because the nutrients carried into them by rivers, estuaries are highly productive ecosystems. They provide important habitats for juvenile stages of many species of fish, shellfish, and crustaceans that are later harvested for food. Indeed, estuaries are sometimes called "nursery" habitats.

Seashores. The seashore biome is formed where the land meets the ocean. The specific character of any given seashore biome is determined by factors such as the intensity of wave action, the frequency of major disturbances, and bottom type. In temperate waters, biomes are often characterized by large species of algae, broadly known as seaweed or kelp. In some cases, so-called kelp "forests" can develop, abundant with marine life. In ecosystems characterized by softer bottoms of sand or mud, invertebrates such as mollusks, echinoderms, crustaceans, and marine worms dominate.

Coral reefs. Coral reefs are marine biomes that are unique to tropical seas. They grow in shallow but relatively infertile areas close to land. Corals are small, tropical marine animals that attach themselves to the seabed and form extensive reefs. The physical structure of the reef is provided by the calcium carbonate skeletons of dead corals. Corals live in symbiosis (in union) with algae, and together create a highly efficient system of obtaining and recycling nutrients. For this reason, coral reefs are

A mangrove swamp in Florida's Everglades National Park. *(Reproduced by permission of The National Parks Service.)*

highly productive, even though they occur in nutrient-poor waters, and support a great variety of species, including living corals, algae, invertebrates, and fishes.

Human-dominated biomes

Urban-industrial techno-ecosystems. The urban-industrial techno-ecosystem consists of a large metropolitan district that is dominated by humans, human dwellings, businesses, factories, other types of buildings, and roads. This biome supports many species in addition to humans. With few exceptions, however, these species are nonnative plants and animals that have been introduced from other places. These organisms typically cannot live independently outside of this biome, unless they are returned to their native biome.

Rural techno-ecosystems. This anthropogenic biome occurs outside of intensively built-up areas. This biome is made up of transportation corridors (such as highways, railways, electric power line transmission corridors, and aqueducts), small towns, and industries involved in the extraction, processing, and manufacturing of products from natural resources (such as mining). Typically, this biome supports mixtures of introduced species and those native species that are tolerant of the disturbances and other stress associated with human activities.

Agroecosystems. Agroecosystems are biomes consisting of regions that are managed and harvested for human use. Farms and ranches are examples of agroecosystems. Many agroecosystems are monocultural, consisting of single types of crops, such as corn or wheat, and are not favorable to native wildlife. The objective in such systems is to manage the species in such a way as to produce a maximum dollar profit. Competing species (weeds and insect pests) are destroyed or prevented from growing or surviving. Less-intensively managed agroecosystems may contain mixtures of species, a form of land management known as polyculture. Polyculture systems may provide habitat for some native wildlife species.

[*See also* **Corals; Desert; Ecosystem; Forests; Lake; Ocean; Rain forest**]

Biophysics

Biophysics is the application of the principles of physics (the science that deals with matter and energy) to explain and explore the form and func-

Words to Know

Computerized axial tomography (CAT scan): An X-ray technique in which a three-dimensional image of a body part is put together by computer using a series of X-ray pictures taken from different angles along a straight line.

Electron microscope: A microscope that uses a beam of electrons to produce an image at very high magnification.

Laser: A device that uses the movement of atoms and molecules to produce intense light with a precisely defined wavelength.

Magnetic resonance imaging (MRI): A technique for producing computerized three-dimensional images of tissues inside the body using radio waves.

Positron-emission tomography: A technique that involves the injection of radioactive dye into the body to produce three-dimensional images of the internal tissues or organs being studied.

Ultracentrifuge: A machine that spins at an extremely high rate of speed and that is used to separate tiny particles out of solution, especially to determine their size.

X ray: A form of electromagnetic radiation with an extremely short wavelength that is produced by bombarding a metallic target with electrons in a vacuum.

X-ray diffraction: A technique for studying a crystal in which X rays directed at it are scattered, with the resulting pattern providing information about the crystal's structure.

tion of living things. The most familiar examples of the role of physics in biology are the use of lenses to correct visual defects and the use of X rays to reveal the structure of bones.

Principles of physics have been used to explain some of the most basic processes in biology such as osmosis, diffusion of gases, and the function of the lens of the eye in focusing light on the retina. (Osmosis is the movement of water across a membrane from a region of higher concentration of water to an area of lower concentration of water. Diffusion of gases is the random motion of gas particles that results in their movement from a region of higher concentration to one of lower concentration.)

The understanding that living organisms obey the laws of physics—just as nonliving systems do—has had a profound influence on the study of biology. The discovery of the relationship between electricity and muscle contraction by Luigi Galvani (1737–1798), an Italian physician, initiated a field of research that continues to give information about the nature of muscle contraction and nerve impulses. Galvani's discovery led to the development of such instruments and devices as the electrocardiograph (to record the electrical impulses that occur during heartbeats), electroencephalograph (to record brain waves), and cardiac pacemaker (to maintain normal heart rhythm).

Medical technology in particular has benefited from the association of physics and biology. Medical imaging with three-dimensional diagnostic techniques such as computerized axial tomography (CAT) scanning, magnetic resonance imaging (MRI), and positron-emission tomography (PET) has permitted researchers to look inside living things without disrupting life processes. Today, lasers and X rays are used routinely in medical treatments.

The use of a wide array of instruments and techniques in biological studies has been advanced by discoveries in physics, especially electronics. This has helped biology to change from a science that *describes* the vital processes of organisms to one that *analyzes* them. For example, one of the most important events of this century—determining the structure of the DNA molecule—was accomplished using X-ray diffraction. This technique has also been used to determine the structure of hemoglobin, viruses, and a variety of other biological molecules and microorganisms.

The ability to apply information discovered in physics to the study of living things led to the development of the electron microscope and ultracentrifuge, instruments that have revealed important information about cell structure and function. Other applications include the use of heat and pressure sensors to obtain information about bodily functions under a variety of conditions. This application of the principles of physics to biology has been of great value in the space program.

[*See also* **Laser; Physics; X rays**]

Biosphere

The biosphere is the space on or near Earth's surface that contains and supports living organisms. It is subdivided into the lithosphere, atmosphere, and hydrosphere. The lithosphere is Earth's surrounding layer, composed

Words to Know

Decomposition: The breakdown of complex molecules—molecules of which dead organisms are composed—into simple nutrients that can be reutilized by living organisms.

Energy: Power that can be used to perform work, such as solar energy.

Global warming: Warming of the atmosphere that results from an increase in the concentration of gases that store heat, such as carbon dioxide.

Nutrient: Molecules that organisms obtain from their environment; they are used for growth, energy, and various other cellular processes.

Nutrient cycle: The cycling of biologically important elements from one molecular form to another and back to the original form.

Photosynthesis: Process in which plants capture light energy from the Sun and use it to convert carbon dioxide and water into oxygen and organic molecules.

Respiration: Chemical reaction between organic molecules and oxygen that produces carbon dioxide, water, and energy.

of solids such as soil and rock; it is about 80 to 100 kilometers (50 to 60 miles) thick. The atmosphere is the surrounding thin layer of gas. The hydrosphere refers to liquid environments such as lakes and oceans that lie between the lithosphere and atmosphere. The biosphere's creation and continuous existence results from chemical, biological, and physical processes.

Requirements for life

For organisms to live, certain environmental conditions (such as proper temperature and moisture) must exist, and the organisms must be supplied with energy and nutrients (food). All the animal and mineral nutrients necessary for life are contained within Earth's biosphere. Nutrients contained in dead organisms or waste products of living cells are transformed back into compounds that other organisms can reuse as food. This recycling of nutrients is necessary because there is no source of food outside the biosphere.

Energy is needed for the functions that organisms perform, such as growth, movement, waste removal, and reproduction. It is the only requirement for life that is supplied from a source outside the biosphere. This energy is received from the Sun. Plants capture sunlight and use it to convert carbon dioxide and water into organic molecules, or food, in a process called photosynthesis. Plants and some microorganisms are the only organisms that can produce their own food. Other organisms, including humans, rely on plants for their energy needs.

The major elements or chemical building blocks that make up all living organisms are carbon, oxygen, nitrogen, phosphorus, and sulfur. Organisms are able to acquire these elements only if they occur in usable

The Beni Biosphere Reserve in Bolivia. *(Reproduced by permission of Conservation International.)*

chemical forms as nutrients. In a process called the nutrient cycle, the elements are transformed from one chemical form to another and then back to the original form. For example, carbon dioxide is removed from the air by plants and incorporated into organic compounds (such as carbohydrates) by photosynthesis. Carbon dioxide is returned to the atmosphere when plants and animals break down organic molecules (a process known as respiration) and when microorganisms break down wastes and tissue from dead organisms (a process known as decomposition).

Evolution of the biosphere

During Earth's long history, life-forms have drastically altered the chemical composition of the biosphere. At the same time, the biosphere's chemical composition has influenced which life-forms inhabit Earth. In the past, the rate at which nutrients were transformed from one chemical form to another did not always equal their transformation back to their original form. This has resulted in a change in the relative concentrations of chemicals such as carbon dioxide and oxygen in the biosphere. The decrease in carbon dioxide and increase in atmospheric oxygen that occurred over time was due to photosynthesis occurring at a faster rate than respiration. The carbon that was present in the atmosphere as carbon dioxide now lies in fossil fuel deposits and limestone rock.

Scientists believe that the increase in atmospheric oxygen concentration influenced the evolution of life. It was not until oxygen reached high concentrations such as exist on Earth today that multicellular organisms like ourselves could have evolved. We require high oxygen concentrations to accommodate our high respiration rates and would not be able to survive had the biosphere not been altered by the organisms that came before us.

Current developments

Most research on the biosphere is to determine the effect that human activities have on the environment—especially on nutrient cycles. Application of fertilizers increases the amount of nitrogen, phosphorus, and other nutrients that organisms can use for growth. These excess nutrients damage lakes, causing overgrowth of algae and killing fish. Fuel consumption and land clearing increase carbon dioxide levels in the atmosphere and may cause global warming (a gradual increase in Earth's temperature) as a result of carbon dioxide's excellent ability to trap heat.

Biosphere 2. Interest in long-term, manned space exploration has also generated research into the development of artificial biospheres. Extended

Gaia Hypothesis

The Gaia hypothesis (pronounced GAY-a), named for the Greek Earth goddess Gaea, is a recent and controversial theory that views Earth as an integrated, living organism rather than as a mere physical object in space. The Gaia hypothesis suggests that all organisms and their environments (making up the biosphere) work together to maintain physical and chemical conditions on Earth that promote and sustain life. According to the hypothesis, organisms interact with the environment as a homeostatic (balancing) mechanism for regulating such conditions as the concentrations of atmospheric oxygen and carbon dioxide. This system helps to maintain conditions within a range that is satisfactory for life. Although scientists agree that organisms and the environment have an influence on each other, there is little support within the scientific community for the notion that Earth is an integrated system capable of regulating conditions to sustain itself. The Gaia hypothesis is a useful concept, however, because it emphasizes the relationship between organisms and the environment and the effect that human activities have on them.

missions into space require that nutrients be cycled in a volume no larger than a building. The Biosphere 2 Project, which received a great deal of popular attention in the early 1990s, provided insight into the difficulty of managing such small, artificial biospheres. The idea behind the project was to establish a planet in miniature where the inhabitants not only survived but learned to live cooperatively and happily together. This is quite revealing, given that human civilization has found it difficult to manage sustainably the much larger biosphere of planet Earth.

One of the most spectacular structures ever built, Biosphere 2 is located in the Sonoran Desert at the foot of the Santa Catalina Mountains not far from Tucson, Arizona. It is the world's largest greenhouse, made of tubular steel and glass, covering an area of three football fields—137,416 square feet (12,766 square meters)—and rising to a height of 85 feet (26 meters) above the desert floor. Within the structure, there is a human habitat and a farm for the Biospherians or inhabitants to work to provide their own food. There are five other wild habitats or biomes representing a savannah, a rain forest, a marsh, a desert, and an ocean. Biosphere 2 is completely sealed so no air or moisture can flow in or out. Nearby are two balloon-like structures that operate like a pair of lungs for Biosphere 2 by maintaining air pressure inside. Only sunlight and electricity are provided from outside.

On September 26, 1991, four women and four men from three different countries entered the Biosphere 2 and the doors were sealed for the two-year-long initial program of survival and experimentation. During this time, the Biospherians attempted to run the farm and grow their own food in the company of some pigs, goats, and many chickens. They shared the other biomes with over 3,800 species of animals and plants that were native to those habitats. The resident scientists observed the interactions of plants and animals, their reactions to change, and their unique methods of living. The Biospherians also had the assignment of experimenting with new methods of cleaning air and water.

On September 26, 1993, the Biospherians emerged from Biosphere 2. It had been the longest period on record that humans had lived in an "isolated confined environment." Unfortunately, the experiment did not live up to expectations. The Biospherians experienced many difficulties, including an unusually cloudy year in the Arizona desert that stunted food crops, rapid growth and expansion of some ant species, and unusual behavior of bees fooled by the glass walls of the structure. In 1996, Columbia University took over operation of the facility, opening a visitors' center later that year. Biosphere 2 has been maintained for study but without human inhabitants. Its future remains uncertain.

[*See also* **Atmosphere, composition and structure; Gaia hypothesis; Photosynthesis; Respiration**]

Biotechnology

Biotechnology is the application of biological processes in the development of products. These products may be organisms, cells, parts of a cell, or chemicals for use in medicine, biology, or industry.

History of biotechnology

Biotechnology has been used by humans for thousands of years in the production of beer and wine. In a process called fermentation, microorganisms such as yeasts and bacteria are mixed with natural products that the microorganisms use as food. In winemaking, yeasts live on the sugars found in grape juice. They digest these sugars and produce two new products: alcohol and carbon dioxide.

Early in the twentieth century, scientists used bacteria to break down, or decompose, organic matter in sewage, thus providing a means for deal-

Words to Know

DNA (deoxyribonucleic acid): A nucleic acid molecule (an organic molecule made of alternating sugar and phosphate groups connected to nitrogen-rich bases) containing genetic information and located in the nucleus of cells.

Hybridization: The production of offspring from two parents (such as plants, animals, or cells) of different breeds, species, or varieties.

Monoclonal antibody: An antibody produced in the laboratory from a single cell formed by the union of a cancer cell with an animal cell.

Recombinant DNA research (rDNA research): A technique for adding new instructions to the DNA of a host cell by combining genes from two different sources.

ing efficiently with these materials in solid waste. Microorganisms were also used to produce various substances in the laboratory.

Hybridization

Hybridization—the production of offspring from two animals or plants of different breeds, varieties, or species—is a form of biotechnology that does not depend on microorganisms. Farmers long ago learned that they could produce offspring with certain characteristics by carefully selecting the parents. In some cases, entirely new animal forms were created that do not occur in nature. An example is the mule, a hybrid of a horse and a donkey.

Hybridization has also been used for centuries in agriculture. Most of the fruits and vegetables in our diet today have been changed by long decades of plant crossbreeding. Modern methods of hybridization have contributed to the production of new food crops and resulted in a dramatic increase in food production.

Discovery of DNA leads to genetic engineering

The discovery of the role of deoxyribonucleic acid (DNA) in living organisms greatly changed the nature of biotechnology in the second half of the twentieth century. DNA, located in the nucleus of cells, is a complex molecule that stores and transmits genetic information. This information provides cells with the directions to carry out vital bodily functions.

With the knowledge of how genetic information is stored and transmitted, scientists have developed the ability to alter DNA, creating new instructions that direct cells to produce new substances or perform new functions. The process of DNA alteration is known as genetic engineering. Genetic engineering often involves combining the DNA from two different organisms, a technique referred to as recombinant DNA research.

There is little doubt that genetic engineering is the best known form of biotechnology today, with animal cloning and the Human Genome Project making headlines in the news. Indeed, it is easy to confuse the two terms. However, they differ in the respect that genetic engineering is only one type of biotechnology.

Monoclonal antibodies

Another development in biotechnology is the discovery of monoclonal antibodies. Monoclonal antibodies are antibodies produced in the laboratory by a single cell. The single cell is formed by the union of two other cells—a cancer cell and an animal cell that makes a particular antibody. The hybrid cell multiplies rapidly, making clones of itself and

An animal cell being microinjected with foreign genetic material. *(Reproduced by permission of Photo Researchers, Inc.)*

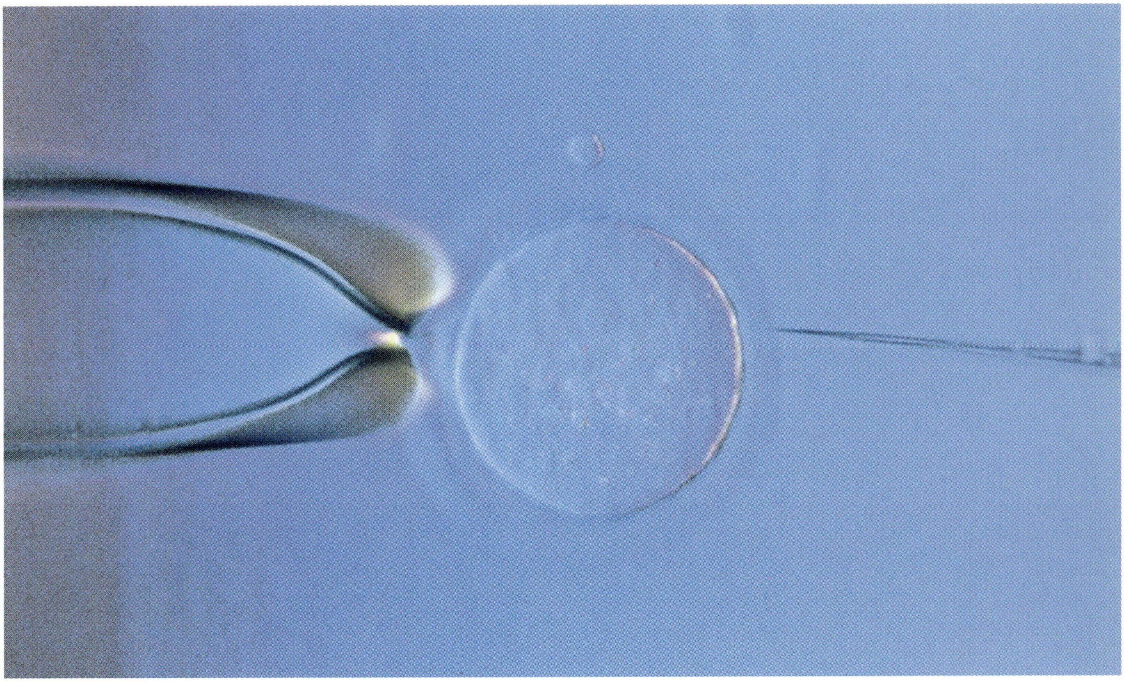

producing large quantities of the antibody. (Antibodies are chemicals produced in the body that fight against foreign substances, such as bacteria and viruses.) Monoclonal antibodies are used in research, medical testing, and for the treatment of specific diseases.

[*See also* **Clone and cloning; Endangered species; Fermentation; Genetic engineering; Human Genome Project; Nucleic acid**]

Birds

Birds are warm-blooded vertebrate (having a backbone) animals whose bodies are covered with feathers and whose forelimbs are modified into wings. Most can fly. Birds are in the class Aves, which contains over 9,500 species divided among 31 living orders. One order, the Passeriformes or perching birds, accounts for more than one-half of all living species of birds.

Most scientists believe that birds evolved from saurischian dinosaurs about 145 million years ago. The first truly birdlike animal, they point out, was *Archaeopteryx lithographica,* which lived during the Jurassic period. Fossils from this animal were found in Germany in the nineteenth century. This 3-foot (1-meter) long animal is considered to be an evolutionary link between the birds and the dinosaurs. It had teeth and other dinosaurian characteristics, but it also had a feathered body and could fly.

A fossil discovery by scientists in 2000, however, threw into doubt the theory of birds' evolution. The fossils in question were excavated in 1969 in Kyrgyzstan, a former Soviet republic, but were not correctly identified until some thirty years later. The animal, *Longisquama insignis,* lived in Central Asia 220 million years ago, not long after the time of the first dinosaurs. From impressions left in stone, it had four legs and what appeared to be feathers on its body. Scientists who analyzed the fossils said the animal had a wishbone virtually identical to *Archaeopteryx* and similar to modern birds. It was a small reptile that probably glided among the trees 75 million years before the earliest known bird. Some scientists believe this challenges the widely held theory that birds evolved from dinosaurs.

Modern characteristics

The bodies of birds are covered with specialized structures known as feathers that grow out of the skin. No other animal has them. Feathers act as a barrier against water and heat loss, are light but very strong, and

Words to Know

Barb: The branches of a feather that grow out of the quill and are held together by barbules in flying birds.

Barbules: Hooks that hold the barbs of a feather together in flying birds.

Bill: The jaws of a bird and their horny covering.

Feathers: Light outgrowths of the skin of birds that cover and protect the body, provide coloration, and aid in flight.

Keel: The ridge on the breastbone of a flying bird to which the flying muscles are attached.

Quill: The hollow central shaft of a feather from which the barbs grow.

provide a smooth, flat surface for pushing against the air during flight. The feathers of most species have color, often bright and beautifully patterned, that serves as camouflage and is used in courtship displays by males.

The modified forelimbs, or wings, of birds are used for flying or gliding. The hind limbs are used for walking, perching, or swimming. Swimming birds typically have webbed feet that aid them in moving through water. The bones of the flying birds are structured for flight. They are very light and have many hollow regions. The wing bones are connected by strong muscles to the keeled, or ridged, breastbone, and the pelvic bones are fused so that they are rigid in flight.

The jaws of birds are modified into a horny beak, or bill, that has no teeth and that is shaped according to the eating habits of each species. Like mammals, birds have a four-chambered heart that pumps blood to the lungs to receive oxygen and then to the body tissues to distribute that oxygen. Fertilization occurs internally, and the female lays hard-shelled eggs—usually in some type of nest—that have a distinct yolk. One or sometimes both parents sit on the eggs until they hatch, and the young of almost all species are cared for by both parents.

The keen eyesight and sensitive hearing of birds aid them in locating food. This is important because their high level of activity requires that they eat often. Birds are also very vocal, using various calls to warn of danger, defend their territory, and communicate with others of their species. Songbirds are any birds that sing musically. Usually, only the male of the species sings. The frequency and intensity of their song is greatest during the breeding season, when the male is establishing a territory and trying to attract a mate.

Birds are found the world over in many different habitats. They range in size from the smallest hummingbird, at less than 2 inches (6 centimeters), to the largest ostrich, which may reach a height of 8 feet (2.4 meters) and weigh as much as 400 pounds (182 kilograms). Many species of birds migrate hundreds or even thousands of miles south every autumn to feed in warmer climates, returning north in the spring.

Flightless birds

Flightless birds lack the keel (high ridge) on the breastbone to which the flight muscles of flying birds are attached. Instead, the breastbone is shaped like a turtle's shell. It has also been described as a raft, giving this group of birds its name, Ratitae (from the Latin *ratis,* meaning "raft"). Ratites have heavy, solid bones and include the largest living birds, such as the ostriches of Africa and the emus of Australia. Kiwis, another type of flightless bird, live in New Zealand and are about the size

A yellow-crowned night heron at the Ding Darling National Wildlife Refuge in Florida. *(Reproduced by permission of Field Mark Publications.)*

of chickens. The penguins of Antarctica are also flightless but are not regarded as ratites. Their powerful flight muscles are used for swimming instead of flying.

Ratites are the oldest living birds and are descended from flying birds who lost the ability to fly. The feathers of ratites differ in structure from those of flying birds. They lack barbules—hooked structures that fasten the barbs of the quill together, providing an air-resistant surface during flight. Instead, the strands that grow from the quill separate softly, allowing air through. This softness makes the feathers of many ratites particularly desirable. Ostrich plumes, for example, have long been used as decoration on helmets and hats.

Human impact on birds

Humans have destroyed birds, both intentionally and unintentionally. Two hundred years ago, birds were considered such an inexhaustible resource that wholesale slaughter of then hardly raised a concern. The greatest impact humans have had on birds has been brought about through human expansion (farms, cities, roads, buildings) into their natural habitats. A by-product of industrial development has been widespread environmental pollution. Pesticides, used on farms to rid fields of insects, have accumulated in many places frequented by birds and have been subsequently ingested by them. Oil spills have also taken their toll on bird populations. It is not surprising, then, that many species have disappeared as a result of human activities and encroachment on the natural environment. According to one scientific estimate, 85 species of birds, representing 27 families, have become extinct since 1600.

Birth

Birth, or parturition (pronounced pahr-chuh-RIH-shuhn), in mammals is the process during which a fully developed fetus emerges from the mother's uterus (or womb) by the force of strong, rhythmic muscle contractions. The birth of live offspring is a reproductive feature shared by mammals, some fishes, and certain invertebrates (such as scorpions), as well as some reptiles and amphibians. Animals that give birth to live offspring are called viviparous (pronounced vie-VIP-uh-rus, meaning "live birth").

In contrast to viviparous animals, other animals such as birds and frogs give birth to eggs; these animals are called oviparous (meaning

Words to Know

Amniotic fluid: The fluid in which the fetus is suspended while in the uterus.

Amniotic sac: A thin membrane forming a sac that contains the amniotic fluid.

Cervix: The narrow, bottom end of the uterus; the opening of the uterus.

Fetus: An unborn mammal at the later stages of development. A human embryo is considered a fetus after eight weeks.

Gestation: The period of carrying young in the uterus before birth.

Labor: The strong, rhythmic contractions of the uterus that result in birth.

Placenta: An organ that develops in certain mammals during gestation through which a fetus receives nourishment from the mother.

Umbilical cord: The cord in most mammals that connects the fetus to the placenta.

Uterus (womb): A muscular organ inside the female mammal in which a baby develops.

Viviparous: Animals that give birth to live offspring.

"egg birth"). Still other animals, such as some fish and reptiles, are ovoviviparous, meaning that the young develop in eggs within the mother's body and hatch either before or immediately after emerging from the mother.

Viviparous animals

In viviparous animals, fertilization of the mother's egg with the father's sperm takes place inside the mother's body. Nutrients are passed from the mother to the developing fetus. In certain mammals, such as humans, this transfer of nutrients occurs through an organ called the placenta, which is formed from the embryo and the mother's uterus.

The carrying of young in the uterus is called gestation (pronounced jes-TAY-shun). The length of time between fertilization and birth in viviparous animals is called the gestation period. The gestation period varies, depending on species. In humans, it is about nine months. Elephants have one of the longest gestation periods of all animals, lasting 22 months.

The birth process

The process of birth in humans normally begins at the end of the gestation period with the release of several hormones that stimulate the mother's uterus to contract. Contractions signal the first stage of labor. In order for the fetus to leave the uterus and enter the birth canal (comprised of the cervix and vagina), it must pass all the way through the cervix, the opening of the uterus.

During the first stage of labor, which can last 12 hours or more, the contractions of the uterus move the fetus toward the cervix, causing the cervix to dilate (widen). With dilation, the cervix opens to accommodate the passage of the baby's head. The amniotic sac also usually ruptures, releasing amniotic fluid that streams out of the vagina. (The amniotic sac is a membrane filled with fluid in which the fetus floats while developing in the uterus.)

A human baby being born. *(Reproduced by permission of Photo Researchers, Inc.)*

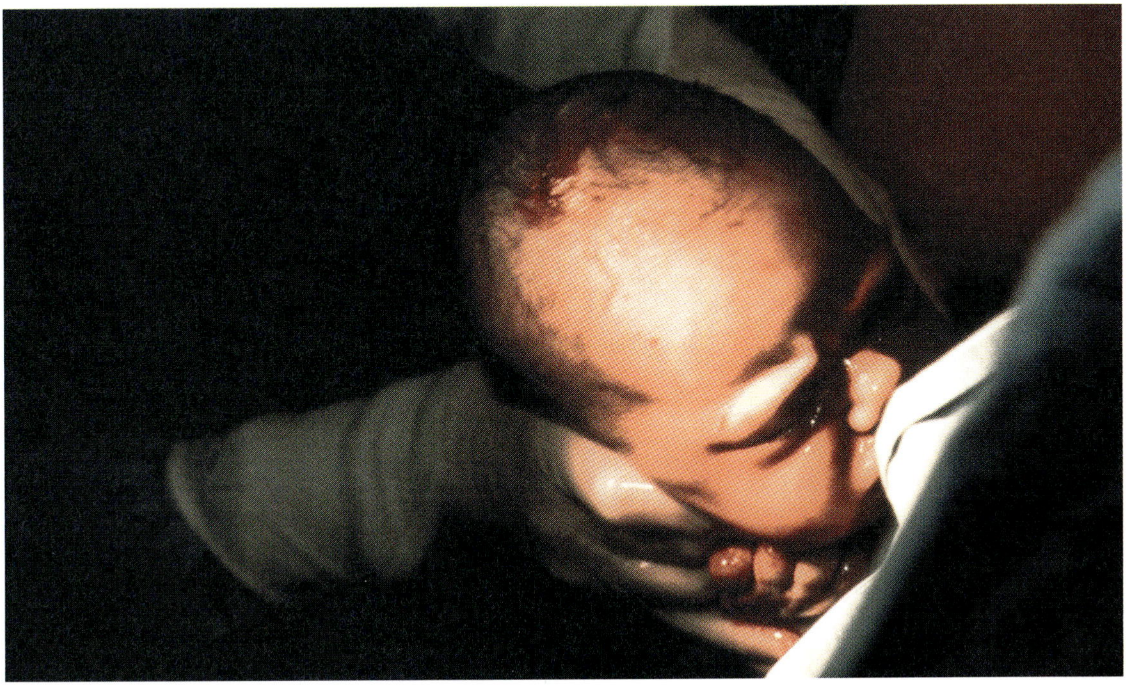

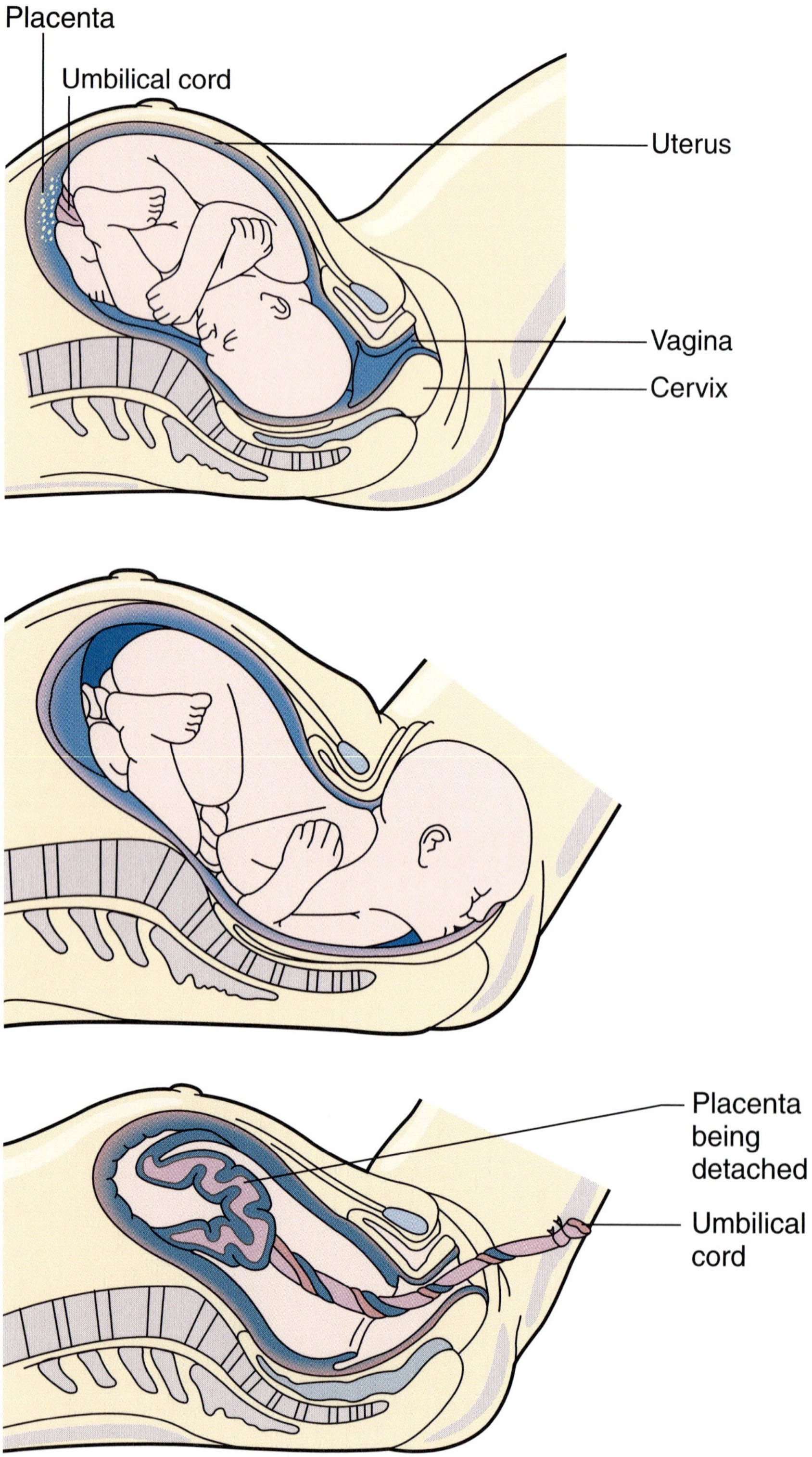

The three stages of birth: dilation of the cervix, expulsion of the fetus, and expulsion of the placenta. *(Reproduced by permission of The Gale Group.)*

During the second stage of labor, lasting anywhere from 30 minutes to 2 hours, the mother uses her abdominal muscles to help push the fetus through and out of the birth canal. In a normal delivery, the baby's head appears first (called crowning) and the rest of the body follows. The umbilical cord that connects the fetus to the placenta is tied and cut. The place on the baby's abdomen where the umbilical cord is attached is the navel, or belly button. The baby is now separated from the mother and must breathe air through its own lungs.

In the third stage of labor, usually lasting from a few minutes to a half hour, contractions cause the placenta and fetal membranes to separate from the wall of the uterus and be expelled from the vagina. The placenta and fetal membranes together are called the afterbirth.

[*See also* **Embryo and embryonic development; Fertilization; Hormones; Reproduction; Reproductive system**]

Birth defects

Birth defects, or congenital defects, are abnormalities that are present at birth. They may be the result of genetic factors such as an inherited disease or a chromosomal abnormality. They may also be caused by environmental factors such as radiation, the mother's use of drugs or alcohol while pregnant, or bacterial or viral infections.

Birth defects can range from very minor, such as a birthmark, to a more serious condition that results in physical deformity or limits the lifespan of the child. Two to three percent of babies born in the United States have a major birth defect.

Chromosomal abnormalities

Some birth defects are caused by chromosomal abnormalities. Chromosomes are structures that are found in the nucleus of every human cell and that contain genes, the units of heredity. Genes are responsible for the physical traits and genetic makeup of an individual. A human fetus inherits 23 chromosomes from its mother and 23 chromosomes from its father, making a total of 46 chromosomes in 23 pairs.

The most common chromosomal abnormalities seen in humans involve an extra chromosome or a missing chromosome. Down syndrome (named for nineteenth-century English physician J. L. H. Down) is a birth disorder caused by the presence of an extra chromosome, so that a

Words to Know

Amniocentesis: Diagnostic technique in which a needle is inserted into the abdomen of a pregnant woman to remove amniotic fluid from the uterus for examination of fetal cells.

Chorionic villus sampling: Diagnostic technique in which a needle is inserted through the abdomen or a thin plastic tube is inserted through the cervix of a pregnant woman to obtain a sample of villi (from the membrane surrounding the fetus) for examination of cells.

Ultrasound: A diagnostic technique that uses sound waves to produce an image of the fetus within the uterus.

fetus with this condition has 47 chromosomes instead of 46. Down syndrome babies have distinctive facial features and other physical abnormalities, are mentally impaired, and often have heart defects. The likelihood of a baby being born with Down syndrome increases with the mother's age.

Inherited disorders

Birth defects that are inherited are caused by abnormal genes. Inherited disorders include sickle-cell anemia, cystic fibrosis, Tay-Sachs disease, and hemophilia.

Sickle-cell anemia is a disease in which the blood cells are shaped like a sickle, or crescent, rather than like a concave circle. This abnormal shape impairs blood flow, resulting in life-threatening complications. Cystic fibrosis is a disease affecting the glands of the body that secrete substances such as sweat or saliva. It is characterized by an overproduction of mucus, leading to severe digestive and respiratory problems. Tay-Sachs disease is a rare, fatal disease caused by the lack of an enzyme (a chemical that speeds up a chemical reaction) that aids in the breakdown of certain fats in the brain. The resulting accumulation of fat deposits in brain cells usually leads to blindness and death by the age of three or four. Hemophilia is an inherited disorder in which failure of the blood to clot results in uncontrolled bleeding.

Physical birth defects

Physical birth defects may result from a combination of genetic and environmental factors that affect the normal development of the fetus. Common physical defects present at birth include clubfoot, in which one or both feet are deformed, and cleft lip and cleft palate, in which a split is present in the lip and roof of the mouth. Spina bifida is a malformation of the spine caused by incomplete closure of the vertebral column (backbone) during development.

Environmental factors

Environmental factors such as the mother's use of tobacco, alcohol, or drugs can affect fetal development and contribute to or cause birth defects. In addition, if the mother is exposed to or infected by an active virus, she may pass it to her child while it is in the womb or during delivery, resulting in complications such as mental retardation in the newborn.

Heavy use of alcohol during pregnancy can result in fetal alcohol syndrome. Babies with this condition typically have low birth weights, physical deformities of the face and head, and varying degrees of mental

An infant with spina bifida. *(Reproduced by permission of Photo Researchers, Inc.)*

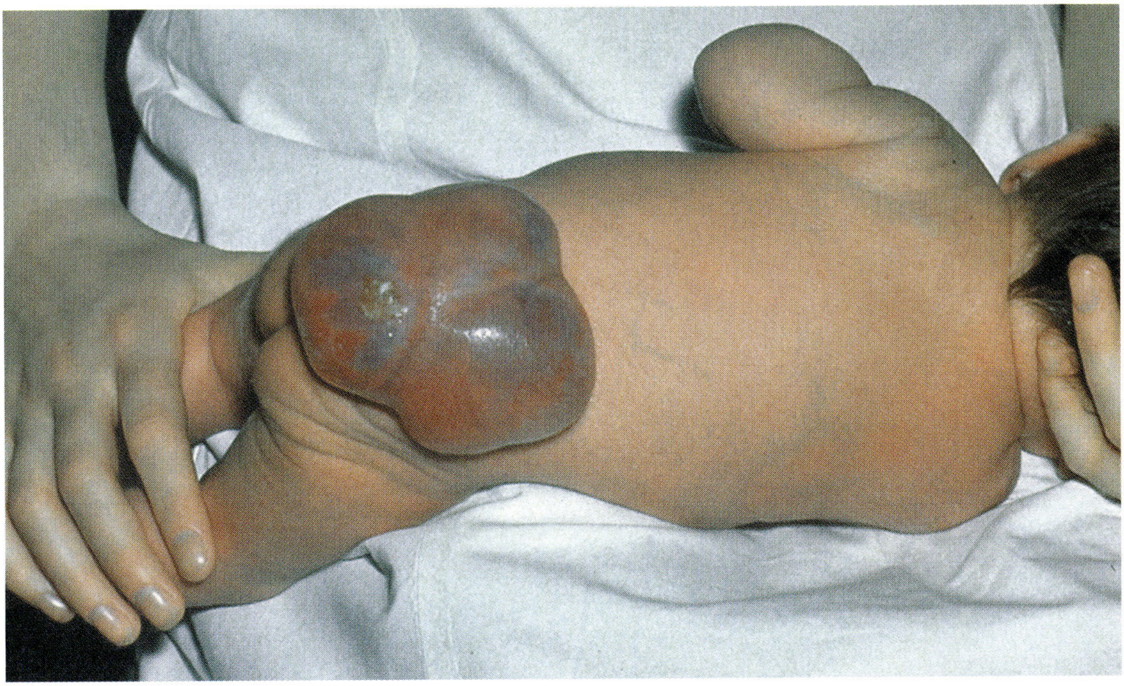

Apgar Score

The Apgar score is the evaluation of a newborn baby's physical condition based on skin color, heart rate, response to stimulation, muscle tone, and respiratory effort. Each criteria is rated from zero to two with a total score of 10 indicating the best possible physical condition. The evaluation determines whether immediate emergency measures are needed. (A low score can indicate possible brain damage.) Because the score is closely related to an infant's life expectancy, it is used as a guideline to advise parents of their baby's chances of survival.

retardation. They may also have behavioral problems and learning difficulties later in life.

Prenatal testing

Prenatal (before birth) testing can detect a number of congenital disorders. Ultrasound, which uses sound waves to produce an image of the fetus, can diagnose spina bifida and certain defects of the heart and other organs. Amniocentesis (pronounced am-nee-oh-sen-TEE-sus) and chorionic (pronounced kor-ee-AH-nik) villus sampling (CVS) are procedures used for detecting genetic disorders such as Down syndrome and sickle-cell anemia. Cells containing the fetus's genetic material are removed from the mother's uterus and tested for the presence of genetic abnormalities.

In recent years, techniques have been developed that allow doctors to operate on the fetus while it is still in the womb. Some congenital defects can be surgically corrected, and blood transfusions can be performed to treat certain conditions detected through prenatal testing.

[*See also* **Birth; Embryo and embryonic development; Genetic disorders**]

Black hole

A black hole, among the most mysterious elements in the universe, is all that remains of a massive star that has used up its nuclear fuel. Lacking energy to combat the force of its own gravity, the star compresses or shrinks in size to a single point, called a singularity. At this point, pressure and density are infinite. Any object or even light that gets too close to a black hole is pulled in, stretched to infinity, and trapped forever. Black holes, so named by American physicist John Wheeler in 1969, are impossible to see, but may account for 90 percent of the content of the universe.

English geologist John Michell and French astronomer Pierre-Simon Laplace first developed the idea of black holes in the eighteenth century. They theorized that if a celestial body were large enough and dense enough, it would exhibit so much gravity that nothing could escape its pull.

This idea can be explained by looking at the effects of gravity on known objects. To break free of Earth's gravity, a spaceship has to travel at a speed of at least 7 miles (11 kilometers) per second. To escape a larger planet like Jupiter, it would have to travel at 37 miles (60 kilometers) per second. And to escape the Sun, it would have to travel at 380 miles (611 kilometers) per second. A large and dense enough object could require the spaceship to go faster than the speed of light, 186,000 miles (299,000 kilometers) per second. However, since nothing can travel faster than the speed of light, nothing would be able to escape the gravity of such an object. Black holes, indeed, are such objects.

Black hole formation

Once a star's nuclear fuel is spent, it will collapse. Without the force of nuclear fusion pushing outward from its core to balance its immense gravity, a star will fall into itself. Average-sized stars, like the Sun, shrink to become white dwarfs (small, extremely dense stars having low brightness) about the size of Earth. Stars up to three times the mass of the Sun explode to produce a supernova. Any remaining matter of such stars ends up as densely packed neutron stars or pulsars (rapidly rotating stars that emit varying radio waves at precise intervals). Stars more than three times the mass of the Sun explode in a supernova and then, in theory, collapse to form a black hole.

When a giant star collapses, its remaining mass becomes so concentrated that it shrinks to an indefinitely small size and its gravity becomes completely overpowering. According to German-born American physicist Albert Einstein's (1879–1955) general theory of relativity, space becomes curved near objects or matter; the more concentrated or dense that matter is, the more space is curved around it. When a black hole forms, space curves so completely around it that only a small opening to the rest of normal space remains. The surface of this opening is called the event horizon, a theorized point of no-return. Any matter that crosses the

Words to Know

Binary star: Pair of stars in a single system that orbit each other, bound together by their mutual gravities.

General theory of relativity: A theory of gravity put forth by Albert Einstein in 1916 that describes gravity as a distortion or curvature of space-time caused by the presence of matter.

Light-year: Distance light travels in one solar year, roughly 5.9 trillion miles (9.5 trillion kilometers).

Pulsars: Rapidly rotating stars that emit varying radio waves at precise intervals; also known as neutron stars because much of the matter within has been compressed into neutrons.

White dwarf: Average-sized star that has collapsed to about the size of Earth and has extreme density and low brightness.

event horizon is drawn in by the black hole's gravity and cannot escape, vanishing across the boundary like water down a drain.

Black hole evidence

Black holes cannot be seen because matter, light, and other forms of energy do not escape from them. They can possibly be detected, however, by their effect on visible objects around them. Scientists believe that as gaseous matter swirls in a whirlpool before plunging into a black hole, that heated matter emits fluctuating X rays. Discovery of such a condition in space, therefore, may indicate the existence of a black hole near the source of those X rays.

In 1971, an X-ray telescope aboard the satellite *Uhuru* detected the first serious black hole candidate in our galaxy, the Milky Way. A black hole is believed to be the companion star in a binary star called Cygnus X-1 (a binary star is a pair of stars in a single system that orbit each other, bound together by their mutual gravities). Cygnus X-1 is emitting intense amounts of X rays, possibly as a result of the unseen companion pulling in stellar material from the other star.

In the 1990s, the Hubble Space Telescope provided scientists with evidence that black holes probably exist in nearly all galaxies and in interstellar space between galaxies. The biggest black holes are those at the center of galaxies. In the giant galaxy M87, located in the constellation Virgo, swirling gases around a suspected massive black hole stretch a distance of 500 light-years, or 2,950 trillion miles (4,750 trillion kilometers).

In early 2001, scientists announced that data from the Chandra X-Ray Observatory and the Hubble Space Telescope provided the best direct evidence for the existence of the theorized event horizons. The two Earth-orbiting telescopes both surveyed matter surrounding suspected black holes that eventually disappeared from view. The Chandra telescope observed X-ray emissions that disappeared around six candidate black holes, while the Hubble telescope observed pulses of ultraviolet light from clumps of hot gas as they faded and then disappeared around Cygnus X-1.

Types of black holes

Until very recently, scientists believed there were only two types of black holes. The first kind, stellar black holes, form from the remains

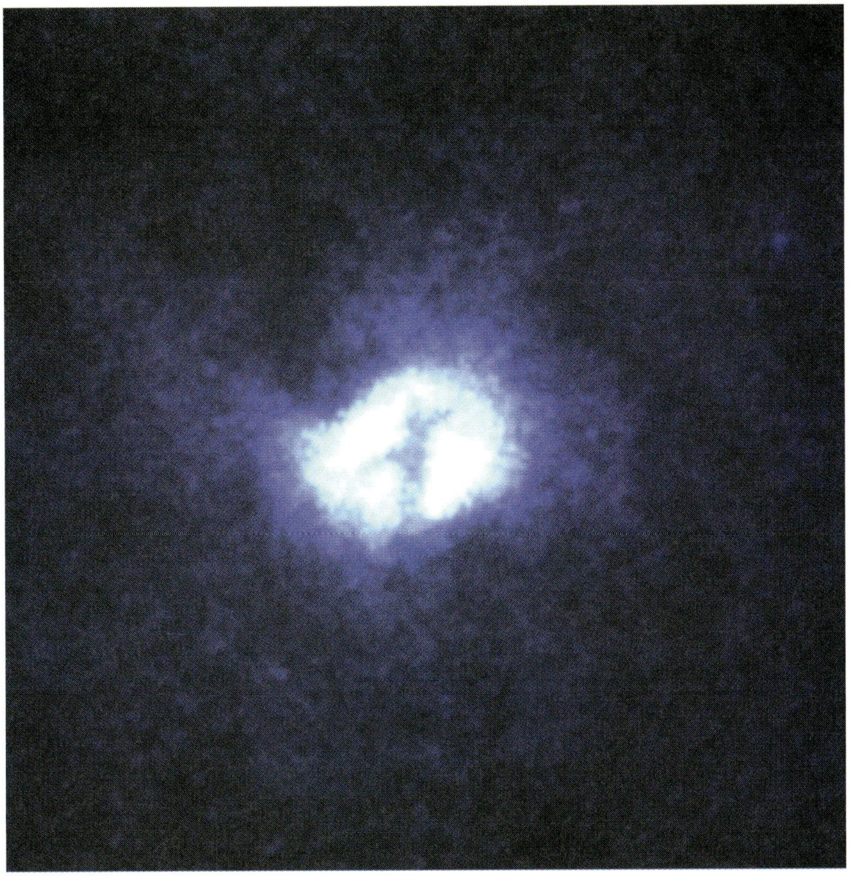

An image of the core of the Whirlpool galaxy M51 (NGC 5149) taken by the Hubble Space Telescope. It shows an immense ring of dust and gas that is thought to surround and hide a giant black hole in the center of the galaxy. *(Reproduced by permission of National Aeronautics and Space Administration.)*

of collapsed stars that have, at most, 10 times the mass of the Sun. The second type, supermassive black holes, are believed to have been formed when the universe was very young. It is also believed they are the most common, existing at the core of every galaxy in the universe. These gigantic black holes have masses up to that of a billion Suns. In 2000, astrophysicists from the Harvard-Smithsonian Center for Astrophysics in Cambridge, Massachusetts, announced they had convincing evidence for a new class of black hole: midsize black holes. Using observations from the Chandra X-Ray Observatory of a galaxy about 12 million light-years from Earth, the scientists theorized the existence of a black hole in that galaxy with a mass of at least 500 Suns. How a midsize black hole like this forms, however, remains a puzzle for scientists.

[*See also* **Relativity, theory of; Star; Supernova**]

Blood

Blood is a fluid connective tissue that performs many functions in the body. It carries oxygen and nutrients to the cells, hormones (chemical messengers) to the tissues, and waste products to organs that remove them from the body. Blood also acts as a defense against foreign microorganisms and helps to keep the body at a constant temperature in warm-blooded animals.

Blood consists of white blood cells, red blood cells, and platelets suspended in plasma, a watery, straw-colored fluid. Plasma makes up about 55 percent of the blood, while blood cells and platelets make up the remaining 45 percent. The average adult human body contains about 6 quarts (approximately 5.6 microliters) of blood.

Plasma

Plasma is made up of 92 percent water, 7 percent proteins, salts, and other substances it transports. Fibrinogen is an important protein involved in blood clotting. Albumins and globulins are proteins that aid in the regulation of fluid in and out of the blood vessels. Proteins called gamma globulins act as antibodies and help protect the body against foreign substances, called antigens.

The salts present in plasma include sodium, potassium, calcium, magnesium, chloride, and bicarbonate. They are involved in many important body functions such as muscle contraction, the transmission of nerve impulses, and regulation of the body's acid-base balance. The re-

Words to Know

Capillary: Microscopic vessels in the tissues that are involved in the exchange of nutrients and other substances between the blood and the tissues.

Clotting factor: A substance that promotes the clotting of blood (stoppage of blood flow).

Erythrocyte: A red blood cell.

Fibrin: A protein in plasma that functions in blood clotting by forming a network of threads that stop the flow of blood.

Hemoglobin: The protein pigment in red blood cells that transports oxygen to the tissues and carbon dioxide from them.

Hemophilia: A genetic disorder in which one or more clotting factors is absent from the blood, resulting in excessive bleeding.

Leukocyte: A white blood cell.

Plasma: The straw-colored liquid portion of the blood that contains water, proteins, salts, nutrients, hormones, and wastes.

Platelet: A disk-shaped cell fragment involved in blood clotting.

Proteins: Large molecules that are essential to the structure and functioning of all living cells.

Red bone marrow: The soft reddish tissue in the cavity of bones from which blood cells are produced.

maining substances in plasma include nutrients, hormones, dissolved gases, and waste products that are being transported to and from body cells. These materials enter and leave the plasma as blood circulates through the body.

Red blood cells

The main function of red blood cells, or erythrocytes (pronounced uh-REE-throw-sites), is the transport of oxygen from the lungs to body tissues. Erythrocytes are tiny disk-shaped structures that are hollowed out on either side. Their small size allows them to squeeze through microscopic blood vessels called capillaries. They number about 5 million per cubic millimeter of blood; in the entire human body, there are about 25 trillion red blood cells.

Red blood cells are formed in the red bone marrow of certain bones, where they produce a substance called hemoglobin. Hemoglobin is a protein pigment that contains iron and that gives red blood cells their color. The hemoglobin in red blood cells combines with oxygen in the lungs, transporting that oxygen to the tissues throughout the body. It also carries carbon dioxide from the tissues back to the lungs, where some of the carbon dioxide is exhaled. Each red blood cell lives only about four months. New red blood cells are constantly being produced in the bone marrow to take the place of old ones.

White blood cells

White blood cells, often called leukocytes (pronounced LUKE-oh-sites), are part of the body's immune system. They defend the body

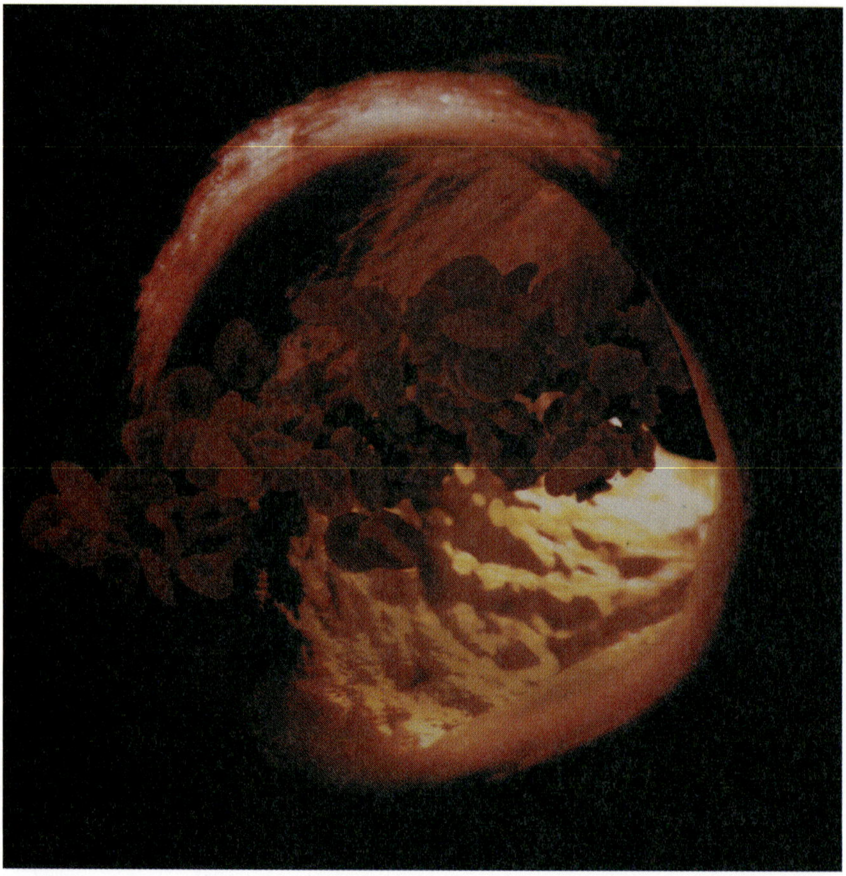

Red blood cells flowing through blood vessels. *(Reproduced by permission of Phototake.)*

against viruses, bacteria, and other invading microorganisms. There are five kinds of white blood cells in human blood: neutrophils, eosinophils, basophils, monocytes, and lymphocytes. Each plays a specific role in the body's immune or defense system. For example, during long-term infections such as tuberculosis (infectious disease of the lungs), monocytes increase in number. During asthma and allergy attacks, eosinophils increase in number.

Lymphocytes make up roughly one-fourth of all white blood cells in the body. They are divided into two classes: T lymphocytes and B lymphocytes. The letter T refers to the thymus, an organ located in the upper chest region where these cells mature. The letter B refers to the bone marrow where these specific lymphocytes mature. T lymphocytes are further divided into four types. Of these four, helper T lymphocytes are the most important. They direct or manage the body's immune response, not only at the site of infection but throughout the body. HIV, the virus that causes acquired immunodeficiency syndrome or AIDS, attacks and kills helper T lymphocytes. The disease cripples the immune system, leaving the body helpless to stave off infections. As AIDS progresses, the number of helper T lymphocytes drops from a normal 1,000 to 0.

All white blood cells are produced in the bone marrow. Some types are carried in the blood, while others travel to different body tissues. There are about 4,000 to 11,000 white blood cells per cubic millimeter of blood in the human body. This number can greatly increase when the body is fighting off infection.

Platelets

Platelets are small, disk-shaped fragments of cells that are broken off from other cells in the bone marrow. They help to control bleeding in a complex process called hemostasis. When an injury to a blood vessel causes bleeding, platelets stick to the ruptured blood vessel and release substances that attract other platelets. Together they form a temporary blood clot. Through a series of chemical reactions, the plasma protein fibrinogen is converted into fibrin. Fibrin molecules form threads that trap red blood cells and platelets, producing a clot that seals the cut blood vessel.

Platelets number about 300,000 per microliter of human blood. They have a short life span, surviving only about 10 days before being replaced.

In an inherited disorder called hemophilia, one or more clotting factors is missing in the blood. Persons with this disorder bleed excessively after injury because their blood does not clot properly.

Artificial blood: Running through the veins of the future?

Since the seventeenth century, doctors have experimented with substitutes for human blood. These substitutes have ranged from milk to oil to the blood from animals. At the beginning of the twenty-first century, with fears of HIV, mad cow disease, and other viruses contaminating the blood supply, the rush to create artificial blood intensified. Artificial or synthetic blood offers many pluses. In addition to helping relieve blood shortages, it could ease doctors' worries about mismatching the blood types of donors and patients. Artificial blood also stays fresher longer than normal blood and does not have to be refrigerated. In theory, artificial blood may be less likely to harbor viruses that infect donated blood. In 2001, having conducted research and testing for many years, several companies in the United States were close to the goal of creating an artificial human blood for use by the medical community.

[*See also* **Circulatory system; Heart; Respiratory system**]

Blood supply

Blood supply refers to the blood resources in blood banks and hospitals that are available for use by the health care community. The blood supply consists of donated blood units (in pints) that are used in blood transfusions. A blood transfusion is a procedure whereby blood is administered through a needle into the vein of a person or animal. Such transfusions are usually performed in order to replace blood lost due to injury or surgery.

Blood banks

Blood banks are institutions that store blood to be distributed to local hospitals and medical centers. There are over 5,000 blood banks in the United States. Together they contain most of the nation's supply of donated blood. Many blood banks are run by the American Red Cross, an organization that also conducts frequent blood drives throughout the country. In fact, the American Red Cross gathers half the blood used in the United States. The blood supply must be replenished constantly to meet the needs of hospitals and trauma units as well as to replace blood components that have a short shelf life.

Words to Know

AIDS (acquired immunodeficiency syndrome): A disease of the immune system believed to be caused by the human immunodeficiency virus (HIV). It is characterized by the destruction of a particular type of white blood cell and susceptibility to infection and other diseases.

American Red Cross: A national organization dedicated to the promotion of public health and the maintenance of the nation's blood supply.

Blood bank: An institution or center that collects, processes, and stores blood for use in blood transfusions.

Blood transfusion: The introduction of blood into a vein of an animal or a human.

Blood type: A classification of blood—A, B, AB, or O—based on the presence or absence of certain antigens on the surface of red blood cells.

Centrifugation: The process of rapidly spinning a solution so that the heavier components will separate from the lighter ones.

HIV (human immunodeficiency virus): The virus believed to cause AIDS.

Rh factor: Certain antigens that are present on the red blood cells of most people and that can stimulate an immune reaction in the bodies of persons who do not have them.

Blood donations

Donation of blood by volunteers is critical in maintaining the supply of blood in blood banks. Beginning in the late 1990s, blood donations in the United States began to increase by 2 to 3 percent per year. But at the same time, the demand for blood increased by 6 to 8 percent. In 2000, about 13 million units of blood were used in the United States.

Blood is collected from a donor by inserting a needle attached to a thin plastic tube into a vein of the arm. Blood flows through the tube and into a sterile plastic bag. The body of an average adult human contains approximately 6 quarts (5.6 microliters) of blood, and the removal of one pint usually has little effect (although some people—especially those with low body weights—may experience temporary dizziness, nausea, or headache). Healthy donors can make blood donations about every eight weeks without causing harm to their bodies.

The collected blood of the donor is tested for hepatitis (a disease of the liver), syphilis (an STD, or sexually transmitted disease), human immunodeficiency virus (HIV; the virus believed to cause AIDS), and related viruses. It is also classified according to blood type and the presence of Rh, or Rhesus, factor. (Blood types are A, B, AB, and O. Rhesus factor is a substance, called an antigen, in the blood of most people.) It is extremely important that blood be marked correctly. Patients receiving donated blood that is incompatible with their own may suffer serious reactions to it. After being collected and classified, whole donated blood is refrigerated.

Separation of blood components

Most donated blood is separated into its components—plasma, red blood cells, white blood cells, and platelets—before being stored. This allows the blood of a single donor to be used for several patients who have different needs. Blood is separated by means of centrifugation, a process in which the blood is rapidly spun so that the heavier blood cells and platelets separate out from the lighter plasma.

Plasma, the liquid part of blood, can be dried into a powder or frozen. Fresh frozen plasma and freeze-dried preparations containing clotting factors are used to treat patients with hemophilia. Hemophilia is an inherited disorder in which certain clotting factors are missing in the blood, resulting in excessive bleeding. Concentrated red blood cells are used to transfuse patients with anemia, a condition in which the blood contains an insufficient number of red blood cells. White blood cells and platelets are used for transfusions in patients who have a deficiency of these components in their blood.

Diseases and the blood supply

When AIDS (acquired immunodeficiency syndrome) began to appear in hemophiliacs and surgical patients in the early 1980s, it was determined that these patients had contracted the disease through donated blood. In 1985, a test was developed to detect HIV—the virus believed to cause AIDS—in blood. Donors are now carefully screened to eliminate any who may be at risk for carrying the AIDS virus. Although the risk of contracting HIV from blood transfusions is remote, some patients who are scheduled to undergo surgery may choose to donate their own blood beforehand, in case a transfusion is necessary.

It is important to know that a blood donor cannot contract AIDS or any other disease by donating blood. The equipment used to collect donated blood is used only once and then discarded. The restrictions on who can donate blood has been expanding steadily to reduce the risk of introducing infections into blood supplies. The American Red Cross runs up to 12 diagnostic tests to screen for viruses and other contaminants before shipping blood. Once-rare procedures, like stripping out white blood cells from donated blood to reduce the side effects from transfusions, is now commonplace.

[*See also* **AIDS (acquired immunodeficiency syndrome); Blood**]

Boolean algebra

Boolean algebra is a form of mathematics developed by English mathematician George Boole (1815–1864). Boole created a system by which certain logical statements can be expressed in mathematical terms. The consequences of those statements can then be discovered by performing mathematical operations on the symbols.

As a simple example, consider the following two statements: "I will be home today" and "I will be home tomorrow." Then let the first statement be represented by the symbol P and the second statement be represented by the symbol Q. The rules of Boolean algebra can be used to find out the consequences of various combinations of these two propositions, P and Q.

In general, the two statements can be combined in one of two ways:

P *or* Q: I will be home today OR I will be home tomorrow.

P *and* Q: I will be home today AND I will be home tomorrow.

Now the question that can be asked is what conclusions can one draw if P and Q are either true (T) or false (F). For example, what conclusion can be drawn if P and Q are both true? In that case, the combination P *or* Q is also true. That is, if the statement "I will be home today" (P) is true, and the statement "I will be home tomorrow" (Q) is also true, then the combined statement, "I will be home today OR I will be home tomorrow" (P or Q) must also be true.

By comparison, suppose that P is true and Q is false. That is, the statement "I will be home today" (P) is true, but the statement "I will be home tomorrow" (Q) is false. Then the combined statement "I will be home today OR I will be home tomorrow" (P or Q) must be false.

Most problems in Boolean algebra are far more complicated than this simple example. Over time, mathematicians have developed sophisticated mathematical techniques for analyzing very complex logical statements.

Applications

Two things about Boolean algebra make it a very important form of mathematics for practical applications. First, statements expressed in everyday language (such as "I will be home today") can be converted into mathematical expressions, such as letters and numbers. Second, those symbols generally have only one of two values. The statements above (P and Q), for example, are either true or false. That means they can be expressed in a binary system: true or false; yes or no; 0 or 1.

Binary mathematics is the number system most often used with computers. Computer systems consist of magnetic cores that can be switched on or switched off. The numbers 0 and 1 are used to represent the two possible states of a magnetic core. Boolean statements can be represented, then, by the numbers 0 and 1 and also by electrical systems that are either on or off. As a result, when engineers design circuitry for personal computers, pocket calculators, compact disc players, cellular telephones, and a host of other electronic products, they apply the principles of Boolean algebra.

[*See also* **Algebra; Computer, digital**]

Botany

Botany is a branch of biology that deals with plant life. It is the study of the structure and the vital processes of plants, including photosynthesis, respiration, and plant nutrition. Among the plants studied are flowering plants, trees, shrubs, and vines. Specialized areas within the field of botany include the study of mosses, algae, lichens, ferns, and fungi.

Divisions of botanical study

Biochemists study the effects of soil, temperature, and light on plants. Plant morphologists study the evolution and development of leaves, roots, and stems, with a special focus on the tissues at various points on stems (called buds) where the cells have the ability to divide. Plant pathologists investigate the causes of plant disease and the effect that pathogens, such as bacteria and fungi, have on forest trees, vegetable

Words to Know

Binomial nomenclature: System of naming plants and animals in which each species is given a two-part name, the first being the genus and the second being the species.

Fossil: Plant or animal remains or impressions from past geologic ages that are preserved in rock.

Gene: A section of a DNA molecule that carries instructions for the formation, functioning, and transmission of specific traits from one generation to another.

Genus: A category of classification made up of species sharing similar characteristics.

Mendelian laws of inheritance: Laws of heredity set forth by Gregor Mendel based on his experiments in breeding pea plants.

Pathogen: A disease-causing organism.

Photosynthesis: Process by which plants capture sunlight and use it to manufacture their own food.

Potato blight: A disease of potatoes caused by a fungus.

Primary producer: Organisms that manufacture their own food from nonliving substances, usually by photosynthesis.

Transpiration: The loss of water from the surfaces of leaves and stems of plants.

crops, grain, and ornamental plants. Economic botanists study the impact of plants as they relate to human needs for food, clothing, and shelter. Plant geneticists study the arrangement and behavior of genes (the physical units of heredity) in plants in order to develop crops that are resistant to diseases and pests. Fossil plants are studied by paleobotanists (pronounced pale-ee-oh-BOT-uh-nists) to determine the earliest appearances of various groups of plants and the conditions under which they existed.

Interdependence

Plants and animals depend on one another for their survival. Plants are primary producers that, through photosynthesis, provide nutrients that animals use to carry out vital body processes. Animals, in turn, contribute to plant distribution, plant pollination, and every other aspect of plant growth and development. Together with zoology (the study of animals), botany is an important aspect of the study of ecology (the interrelationship of living things and their environments).

History of botany

The field of botany began to take form with the work of Greek philosopher Aristotle (384–322 B.C.), the first person to classify plants.

A showy lady-slipper orchid. *(Reproduced by permission of Field Mark Publications.)*

He divided them into categories according to size and appearance. Many years later, Swedish botanist Carolus Linnaeus (1707–1778) contributed greatly to the study of botany by devising a comprehensive classification system for plants that is still used today. In 1753, Linnaeus published his *Species Plantarum,* in which he classified every known species of plant according to its structure and its similarity to other species. He also gave each plant a two-part name (called binomial nomenclature), consisting of the genus (the biological classification between family and species) and a second descriptive word.

The first scientific experiment in plant nutrition was conducted by Belgian physician Jan Baptista van Helmont (1577–1644). In growing a tree using only water as nourishment, van Helmont proved that the soil in which the tree was planted was not the only source of plant nutrients. English physiologist Stephen Hales (1677–1761) studied plant transpiration (loss of water from the surfaces of plant leaves and stems) and is credited with establishing plant physiology as a science.

During the nineteenth century, advances were made in the study of plant diseases, spurred by the potato blight in Ireland in the 1840s. Caused by a fungus that destroyed the entire potato crop, the potato blight resulted in over one million deaths from starvation and led to a mass migration of Irish to America.

The modern science of plant genetics developed from the work of Gregor Mendel (1822–1884), an Austrian botanist and monk. His breeding experiments with pea plants provided information on the nature of genes and their role in the inheritance of characteristics between generations. He formulated the Mendelian laws of inheritance, which were applied after 1900 to plant breeding.

Research in botany includes developing new and hardier species of crops, controlling plant diseases, discovering new medicines from plants, and studying the effects of human intervention (such as pollution and logging) on plant life. Exploring ways of maintaining an ecological balance that continues to sustain both plant and animal life is an important subject of study as well.

[*See also* **Plant**]

Brain

The brain is a mass of nerve tissue located in an animal's head that controls the body's functions. In simple animals, the brain functions like a

Words to Know

Blood-brain barrier: A modification of capillaries in the brain stem that prevent certain chemicals from entering the brain through the bloodstream.

Broca's area: An area in the cerebrum that organizes thought and coordinates muscles for speech.

Ganglion: A structure comprised of nerve cell bodies, usually located outside the central nervous system.

Glial cells: Nerve cells (other than neurons) located in the brain that protect, support, and assist neurons.

Neuron: A type of nerve cell.

Stimulus: Anything that incites an organism to action, such as light, sound, or moisture.

Wernicke's area: An area in the cerebrum that processes information from written and spoken language.

switchboard, picking up signals from sense organs and passing the information to muscles. In more advanced animals, particularly vertebrates, the brain is much more complex and is capable of far more advanced behaviors.

The brain is part of an animal's central nervous system, which receives and transmits impulses. It works with the peripheral nervous system, which carries impulses to and from the brain and spinal cord via nerves that run throughout the body.

Invertebrate brain

The simplest brains are those found in invertebrates, animals that lack a backbone. For example, roundworms have a simple brain and nervous system consisting of approximately 300 nerve cells, or neurons. Sensory neurons located in the head end of the animal detect stimuli from the environment and pass messages to the brain. The brain then sends out impulses through a nerve cord to muscles, which respond to the stimulus. The way that neurons in the brain process the data received determines the response made by the animal.

Somewhat more advanced organisms have more complex nervous systems. A key component of such systems are ganglia, masses of neurons that can take in and process information. The brain of an earthworm, for example, consists of a pair of ganglia at the animal's head end.

Suppose that an earthworm encounters some external stimulus, such as touch, light, or moisture. That stimulus is detected by receptor cells in the skin, which send out a message along a pair of nerves in each of the earthworm's segments. These nerves carry the message to the brain and also to smaller ganglia in each of the worm's segments, where the signals are analyzed. The central nervous system then transmits impulses

A comparison of the brains of an earthworm, an insect, a bird, and a human. *(Reproduced by permission of The Gale Group.)*

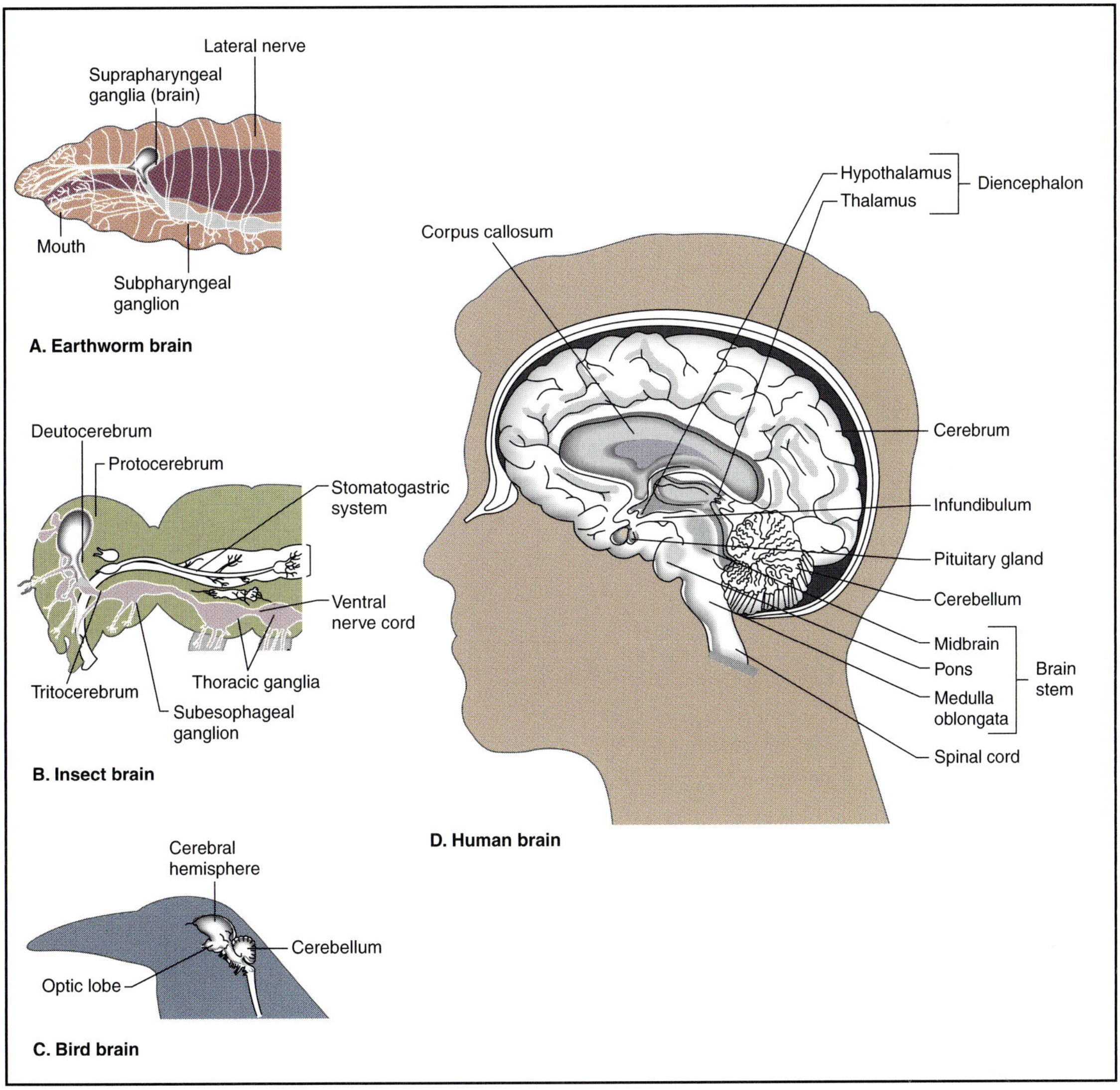

along nerves that coordinate muscle action, causing the earthworm to move toward or away from the stimulus.

Insect brain. In insects, specialized sense organs detect information from the environment and transmit it to the central nervous system. Such sense organs include simple and compound eyes, sound receptors on the thorax (the main body) or in the legs, and taste receptors. The brain of an insect consists of a ganglion in the head. Ganglia are also found in some segments of the insect's body. The information that insects use for behaviors such as walking, flying, mating, and stinging is stored in these segmental ganglia. In experiments in which heads are cut off of cockroaches and flies, these insects continue to learn.

Vertebrate brain

The central nervous system of vertebrates (animals with backbones) consists of a single spinal cord, which runs along the animal's back, and a highly developed brain. The brain is the dominant structure of the nervous system. It is the master controller of all body functions, and the analyzer and interpreter of complex information and behavior patterns. One can think of the brain as a powerful computer that uses nerve cells rather than silicon chips. The peripheral nervous system, composed of nerves which run to all parts of the body, transmits information to and from the central nervous system.

The vertebrate brain is divided into three main divisions: the forebrain, the midbrain, and the hindbrain. The hindbrain connects the brain to the spinal cord, and a portion of it, called the medulla oblongata, controls important body functions such as the breathing rate and the heart rate. Also located in the hindbrain is the cerebellum, which controls balance.

The forebrain consists of the cerebrum, thalamus, and hypothalamus. Among its other functions, the forebrain controls the sense of smell in vertebrates. The midbrain is the location of the optic lobes, responsible for receiving and interpreting visual signals. The midbrain is also the source of an organism's motor responses.

During the first few weeks of development, the brain of a vertebrate looks like a series of bulges in the tube of nerve cells. There is very little difference among early brains of fish, amphibians, reptiles, birds, and mammals. As the brain develops, however, the bulges enlarge. Each type of vertebrate acquires its own specific adult brain that helps it survive in its environment. In the forebrain of fish, for example, the olfactory (smell) sense is well developed, whereas the cerebrum serves merely as a relay station for impulses. In mammals, on the other hand, the olfactory division is included in the system that also controls emotions, and the cerebrum is highly developed, operating as a complex processing center for information. Optic lobes are well developed in the midbrain of nonmammalian vertebrates, whereas in mammals the vision centers are mainly in the forebrain. In addition, a bird's cerebellum is large compared to the rest of its brain, since it controls coordination and balance in flying.

Human brain. The living human brain is a soft, shiny, grayish white, mushroom-shaped structure. Encased within the skull, it is a 3-pound (1.4-kilogram) mass of nerve tissue that keeps us alive and functioning. On average, the brain weighs 13.7 ounces (390 grams) at birth, and by age 15 grows to approximately 46 ounces (1,300 grams). The human brain is composed of up to one trillion nerve cells. One hundred billion of these are neurons, and the remainder are supporting (glial) cells. Neurons receive, process, and transmit impulses, while glial cells (neuroglia) protect, support, and assist neurons.

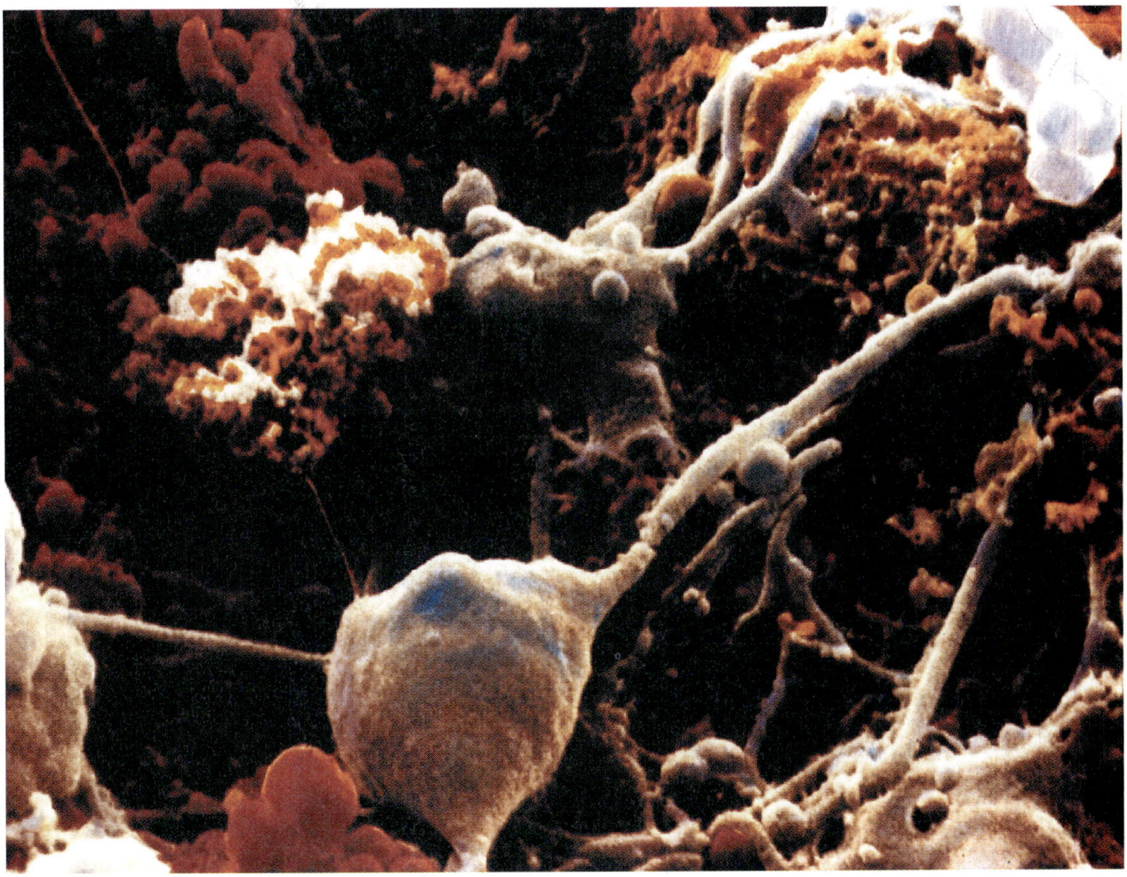

Color scanning electron micrograph of brain cells. The large cells with long, thin branches are neurons. The others are glial cells, specialized structures that support and protect neurons. *(Reproduced by permission of Photo Researchers, Inc.)*

The brain is protected by the skull and by three membranes called the meninges. The outermost membrane is known as the dura mater; the middle as the arachnoid; and the innermost as the pia mater. Also protecting the brain is cerebrospinal fluid, a liquid that circulates between the arachnoid and pia mater. Many bright red arteries and bluish veins on the surface of the brain penetrate inward. Glucose, oxygen, and certain ions pass easily from the blood into the brain, whereas other substances, such as antibiotics, do not. Capillary walls are believed to create a blood-brain barrier that protects the brain from a number of biochemicals circulating in the blood.

The parts of the brain can be studied in terms of structure and function. Four principal sections of the human brain are the brain stem (the hindbrain and midbrain), the diencephalon, the cerebrum, and the cerebellum.

The brain stem. The brain stem is the stalk of the brain, and is a continuation of the spinal cord. It consists of the medulla oblongata, pons, and midbrain. The medulla oblongata is actually a portion of the spinal cord that extends into the brain. All messages that are transmitted between the brain and spinal cord pass through the medulla. Nerves on the right side of the medulla cross to the left side of the brain, and those on the left cross to the right. The result of this arrangement is that each side of the brain controls the opposite side of the body.

Three vital centers in the medulla control heartbeat, rate of breathing, and diameter of the blood vessels. Centers that help coordinate swallowing, vomiting, hiccuping, coughing, and sneezing are also located in the medulla. A region within the medulla helps to maintain the conscious state. The pons (meaning "bridge") conducts messages between the spinal cord and the rest of the brain, and between the different parts of the brain. The midbrain conveys impulses from the cerebral cortex to the pons and spinal cord. It also contains visual and audio reflex centers involving the movement of eyeballs and head.

Twelve pairs of cranial nerves originate in the underside of the brain, mostly from the brain stem. They leave the skull through openings and extend as peripheral nerves to their destinations. Cranial nerves include the olfactory nerve that brings messages about smell from the nose and the optic nerve that conducts visual information from the eyes.

The diencephalon. The diencephalon lies above the brain stem, and includes the thalamus and hypothalamus. The thalamus is an important relay station for sensory information coming to the cerebral cortex from other parts of the brain. The thalamus also interprets sensations of pain, pressure, temperature, and touch, and is concerned with some of our emotions and memory. It receives information from the environment in the form of sound, smell, and taste.

The hypothalamus performs numerous important functions. These include the control of the autonomic nervous system. The autonomic nervous system is a branch of the nervous system involved with control of a number of body functions, such as heart rate and digestion. The hypothalamus helps regulate the endocrine system (which produces hormones, chemical messengers that regulate body functions) and controls normal body temperature. It tells us when we are hungry, full, and thirsty. It helps regulate sleep and wakefulness, and is involved when we feel angry and aggressive.

The cerebrum. The cerebrum makes up about 80 percent of the brain's weight. It lies above the diencephalon. The cerebral cortex is the outer layer of the brain and is made up of a material known as gray matter, consisting of many nerve cell bodies. The tissue of the cerebral cortex is about 0.08 to 0.16 inches (2 to 4 millimeters) thick, and if spread out would have a surface area of about 5 square feet (1.5 square meters), about one-half the size of an office desk. White matter, composed of nerve fibers covered with a fatty-like coating known as myelin sheaths, lies beneath the gray matter.

A deep fissure separates the cerebrum into a left and right hemisphere (halves). Each cerebral hemisphere is divided into regions known as frontal, temporal, parietal, and occipital lobes. The corpus callosum, a large bundle of fibers, connects the two cerebral hemispheres.

The cerebral cortex is the portion of the brain that provides the most important distinctions between humans and other animals. It is responsible for the vast majority of functions that define what we mean by "being human." It enables us not only to receive and interpret all kinds of sensory information, such as color, odor, taste, and sound, but also to remember, analyze, interpret, make decisions, and perform a host of other "higher" brain functions.

By studying animals and humans who have suffered damage to the cerebral cortex, scientists have found that various parts of this region have specific functions. For example, spoken and written language are transmitted to a part of the cerebrum called Wernicke's area, where meaning is extracted. Instructions are then sent to Broca's area, which controls the movement of muscles throughout the body. Within Broca's area, thoughts are translated into speech and muscles are coordinated for speaking. Impulses from other motor areas direct our hand muscles when we write and our eye muscles when we scan the page for information.

Memory

One of the most fascinating of all brain functions is memory. Memory refers to the brain's ability to recall events that have taken place at some time in the past. Scientists have learned that two kinds of memory exist, short-term and long-term memory. They believe that the way in which these two kinds of memory function are somewhat different from each other. People who have a condition known as retrograde amnesia, for example, lose the ability to remember events that occurred immediately before some kind of shock, such as a blow to the head. Yet, they can easily remember events that occurred days, weeks, months or years before that shock.

Scientists are still uncertain as to how the brain remembers things. They use the term memory trace to describe changes in the brain that correspond to the creation of memory. But no one really knows exactly what a memory trace corresponds to in terms of brain structure, chemistry, or function.

According to the most popular current theory of memory, exposure to stimuli can cause changes in the connections that neurons make with each other. These changes may be the "memory traces" that scientists talk about. Those neural connections appear to be able to survive for very long periods of time and can be recalled when a person decides to recall them or when some stimulus causes them to reappear.

Some exciting research on memory has suggested that nerve cells may actually grow and change as they are exposed to light, sounds, chemicals, and other stimuli. The new patterns they form may in some way be connected to the development of a memory trace in the brain.

Association areas of the cerebrum are concerned with emotions and intellectual processes, by connecting sensory and motor functions. In our association areas, innumerable impulses are processed that result in memory, emotions, judgment, personality, and intelligence.

Two surprisingly small areas at the front of the cerebrum, located on each hemisphere roughly above the outer edge of the eyebrow, are the brain's centers for high-level thinking. In 2000, scientists announced that this paired region, called the lateral prefrontal cortex, was activated in people who were given tests involving verbal and spatial problems. This is the same region of the brain that previous research studies had shown to be important for solving novel tasks, keeping many things in mind at once, and screening out irrelevant or unimportant information. Scientists also believe the lateral prefrontal cortex acts as a global workspace for organizing and coordinating information and carrying it back to other parts of the brain as needed.

Certain structures in the cerebrum and diencephalon make up the limbic system. These regions are responsible for memory and emotions, and are associated with pain and pleasure.

By studying patients whose corpus callosum had been destroyed, scientists have learned that differences exist between the left and right sides of the cerebral cortex. The left side of the brain functions mainly in speech, logic, writing, and arithmetic. The right side of the brain, on the other hand, is more concerned with imagination, art, symbols, and spatial relations.

The cerebellum. The cerebellum is located below the cerebrum and behind the brain stem, and is shaped like a butterfly. The "wings" are the cerebellar hemispheres, and each consists of lobes that have distinct grooves or fissures. The cerebellum controls the movements of our muscular system needed for balance, posture, and maintaining posture.

Brain disorders

As with any other part of the body, the brain is subject to a variety of disfunctions and disorders. Four of the most common of these are coma, epilepsy, migraine, and stroke.

Coma. The term coma comes from the Greek word *koma,* meaning "deep sleep." Medically, coma is a state of extreme unresponsiveness in which an individual exhibits no voluntary movement or behavior. In a deep coma, stimuli, even painful stimuli, are unable to effect any response. Normal reflexes may be lost.

The term coma is used for a side variety of conditions ranging from drowsiness or numbness at the least extreme to brain death at the worst extreme.

Coma is the result of something that interferes with the functioning of the cerebral cortex and/or the functioning of the structures that make up the reticular activating system (RAS). The RAS is a network of structures (including the brain stem, the medulla, and the thalamus) that work together to control a person's tendency to remain awake and alert.

A large number of conditions can result in coma, including damage to the brain itself, such as brain tumors, infections, and head injuries. They

Split-brain Research

The two hemispheres of the human brain are not created equal. Scientists have suspected for centuries that each hemisphere of the brain has specialized functions. As early as the 1860s, French physician Paul Broca (1824–1880) showed that patients with speech problems had damage to the left side of their brains.

Additional research on split-brain functions later came from many other sources. For example, some patients with severe epilepsy have had the corpus callosum in their brain severed to relieve their discomfort. This surgery has, as a by-product, produced information on the way the two hemispheres function. A simpler way to study the two hemispheres is simply to protect one hemisphere from receiving information, such as placing a card over one eye, covering one ear, or providing sensory inputs to only one hand, arm, or foot.

As a result of studies of this kind, scientists have been able to assign certain types of function to one hemisphere of the brain or the other. Perhaps the most obvious of these functions is handedness. In general, about 90 percent of all humans are right-handed, a characteristic that can now be traced to the control of that function in the left hemisphere of most human brains.

Scientists now believe that language functions, such as the ability to speak, read, name objects, and understand spoken language

may also involve changes in the way the brain functions, such as a decrease in the availability of substances necessary for appropriate brain functioning, such as oxygen, glucose, and sodium; the presence of certain substances disrupting the functioning of neurons, such as drugs or alcohol in toxic (poisonous) quantities; or changes in the levels of certain essential brain chemicals due to seizures.

The ultimate results of coma depend on a number of factors. In general, it is extremely important for a physician to determine quickly the cause of a coma, so that potentially reversible conditions are treated immediately. For example, an infection may be treated with antibiotics, a brain tumor may be removed, brain swelling from an injury can be reduced with certain medications.

Outcome from a coma depends on its cause and duration. In drug poisonings, for example, extremely high rates of recovery can be expected, are a function of the left hemisphere. The left hemisphere is also thought to be responsible for numerical and analytical skills. In contrast, the right hemisphere in most humans is thought to control nonverbal activities, such as the ability to draw and copy geometric figures, various musical abilities, visual-spatial reasoning and memory, and the recognition of form using vision and touch.

There is also evidence that the two hemispheres of the brain process information differently. It seems that the right hemisphere tends to process information in a more simultaneous manner, processing and bringing diverse pieces of information together. The left hemisphere seems to process information in a logical and sequential manner, proceeding in a more step-by-step manner than the right hemisphere.

In terms of emotions, there are some very intriguing findings. Some of these findings come from studies of individuals who showed uncontrollable laughter or crying. This evidence suggests that the left hemisphere is highly involved in the expression of positive emotions, while the right hemisphere is highly involved in the expression of negative emotions. Some researchers believe that the two hemispheres of the brain usually inhibit each other so that there is a balance, making uncontrollable emotional outbursts rare.

following prompt medical attention. Patients who have suffered head injuries tend to do better than patients whose coma was caused by other types of medical illnesses. Excluding drug poisoning-induced comas, only about 15 percent of patients who remain in a coma for more than a few hours make a good recovery. Adult patients who remain in a coma for more than four weeks have almost no chance of regaining their previous level of functioning. However, children and young adults have regained functioning after even two months in a coma.

Epilepsy. The term epilepsy is derived from the Greek word for seizure. It describes a condition marked by irregularities in the body's electrical rhythms and is characterized by convulsive attacks (violent involuntary muscle contractions) during which a person may lose consciousness. The outward signs of epilepsy may range from only a slight smacking of the lips or staring into space to a generalized convulsion. It is a condition that can affect anyone, from the very young to adults, of both sexes and any race. Epilepsy was first described by the Greek physician Hippocrates, known as the father of medicine, who lived in the late fifth century B.C.

The number of people who have epilepsy is not known. Some authorities say that up to one-half of 1 percent of the population are epileptic. But other experts believe this estimate to be too low. Many cases of epilepsy, those with very subtle symptoms, are not reported.

The cause of epilepsy remains unknown. However, scientists are often able to determine the area of the brain that is affected by the manner in which the condition is demonstrated. For example, Jacksonian seizures, which are localized twitching of muscles, originate in the frontal lobe of the brain in the motor cortex. A localized numbness or tingling indicates an origin in the parietal lobe on the side of the brain in the sensory cortex.

The recurrent (repeated) symptoms associated with epilepsy, then, are the result of unusually large electrical discharges from neurons in a particular region of the brain. These discharges can be seen on the standard brain test called the electroencephalogram (EEG). For this test, electrodes (devices that conduct electrical current) are applied to specific areas of the head to pick up the electrical waves generated by the brain. If the patient experiences an epileptic episode while he or she is wired to the EEG, the abnormal brain waves can easily be seen and the determination made as to their origin in the brain. If the patient is not experiencing a seizure, however, abnormalities will usually not be found in the EEG.

Perhaps the best known examples of epilepsy known to the general public are grand mal and petit mal. The term *mal* comes from the French word of "illness," while *grand* and *petit* refer respectively to "large" and "small" episodes of the illness. In the case of grand mal, an epileptic is likely to have some indication that a seizure is imminent immediately prior to the seizure. This feeling is called an aura. Very soon after feeling the aura, the person will lapse into unconsciousness and experience generalized muscle contractions that may distort the body position. The thrashing movements of the limbs that follow in a short time are caused by opposing sets of muscles alternating in contractions. The person may also lose control of the bladder and/or bowels. When the seizures cease, usually after three to five minutes, the person may remain unconscious for up to half an hour. Upon awakening, the person may not remember having had a seizure and may be confused for a time.

In contrast to the drama of the grand mal seizure, the petit mal may seem insignificant. The person interrupts whatever he or she is doing and for up to about 30 seconds may show subtle outward signs, such as blinking of the eyes, staring into space, or pausing in conversation. After the seizure has ended, the person resumes his or her previous activity, usually totally unaware of the interruption that took place. Petit mal seizures are associated with heredity, and they never occur in people over the age of 20 years. Oddly, though the seizures may occur several times a day, they do so in most cases when the person is quiet and not during periods of activity. After puberty these seizures may disappear or they may be replaced by the grand mal type of seizure.

Treatment. A number of drugs are available for the treatment of epilepsy. The oldest is phenobarbital, which has the unfortunate side effect of being addictive. Other drugs currently on the market are less addictive, but all have the possibility of causing unpleasant side effects such as drowsiness or nausea or dizziness.

The epileptic person needs to be protected from injuring himself or herself during an attack. For the person having a petit mal seizure, little usually needs to be done. Occasionally these individuals may lose their balance and need to be helped to the ground to avoid hitting their head. Otherwise, they need little attention.

The individual in a grand mal seizure should not be restrained, but may need to have some help to avoid striking his or her limbs or head on the floor or any nearby objects. If possible, the person should be rolled to one side. This action will maintain an open airway for the person to breathe by allowing the tongue to fall to one side.

Epilepsy is a recurrent, lifelong condition that must be managed on a long-term basis. Medication can control seizures in a substantial percentage of people, perhaps up to 85 percent of those with grand mal manifestations. Some people will experience seizures even with maximum dosages of medication. These individuals need to wear an identification bracelet to let others know of their condition. Epilepsy is not a reflection of insanity or mental retardation in any way. In fact, many who experience petit mal seizures are of above-average intelligence.

Migraine. Migraine is a particularly severe form of headache. It was first described during the Mesopotamian era, about 3000 B.C. Migraine is a complex condition that is still poorly understood. The term does not apply to a single medical condition, but is applied to a variety of symptoms that are often numerous and changeable. Migraine sufferers find that their headaches are provoked by a particular stimulus, such as stress, loud noises, missed meals, or eating particular foods such as chocolate or red wine.

A migraine condition can generally be divided into four distinct phases. The first phase is known as the prodrome. Symptoms develop slowly over a 24-hour period preceding the onset of the headache, and often include feelings of heightened or dulled perception, irritability or withdrawal, cravings for certain foods, and other features.

The second phase, known as the aura, features visual disturbances that may be described as flashing lights, shimmering zig-zag lines, spotty vision, and other disturbances in one or both eyes. Other sensory symptoms may occur as well, such as pins and needles or numbness in the hands. All of these symptoms can be acutely distressing to the patient. This phase usually precedes the onset of headache by one hour or less.

Phase three consists of the headache itself, usually described as severe, often with a throbbing or pulsating quality. The pain may occur on one or both sides of the head, and may be accompanied by nausea and vomiting and intolerance of light (photophobia), noise (phonophobia), or movement. This phase may last from 4 to 72 hours. During the final phase, called the postdrome, the person often feels drained and washed-out. This feeling generally subsides within 24 hours.

Migraines appear to involve changes in the patterns of blood circulation and of nerve transmissions in the brain. Scientists currently believe that migraines develop in three phases. The first step takes place in the midbrain. For reasons not fully understood, cells that are otherwise functioning normally in this region begin sending abnormal electrical signals along their projections to other brain centers, including the visual cortex. The second step is activation of the blood vessels in the brain, wherein arteries may contract or dilate (expand). The third step is activation of nerve cells that control the sensation of pain in the head and face. Some patients may experience only one of these three stages. This fact could explain individuals who experience only the aura, without the pain phase, for example.

Some recent research suggests a connection between migraine and levels of serotonin, a neurotransmitter found in the brain and numerous other cells and tissues. Migraine attacks have been correlated with falling levels of serotonin in the body. The connection has been strengthened by the observation that the drug sumatriptan, which closely resembles serotonin chemically, is highly effective in treating migraine.

Stroke. Stroke is a medical condition characterized by the sudden loss of consciousness, sensation, and voluntary movement caused by the loss of blood flow to the brain. Stroke is also called a cerebral vascular accident or CVA. It is caused by one of two events, a ruptured artery or an artery that has become closed off because a blood clot has lodged in it. Stroke resulting from a burst blood vessel is called a hemorrhagic stroke, while one caused by a clot is called a thrombotic stroke. Blood circulation to the area of the brain served by that artery stops at the point of disturbance, and brain tissue beyond that point is damaged by the lack of oxygen and begins to die.

Stroke is the third leading cause of death in the United States after heart attack and all forms of cancer. Approximately 500,000 strokes, new and recurrent, are reported each year. Of these, about 150,000 will be fatal. Today approximately 3,000,000 Americans who have had a stroke are alive.

How a stroke occurs. The brain requires a constant and steady flow of blood in order to carry out its functions. Blood delivers the oxygen and nutrients needed by the brain cells. If this blood flow is interrupted for any period of time and for any reason, brain cells begin to die quickly.

A burst blood vessel may occur in a weak area in the artery, or a blood vessel that becomes plugged by a floating blood clot. In either case, blood is no longer supplied to brain tissue beyond the point of the occurrence. The effect of the interruption in circulation to the brain depends upon the area of the brain that is involved. Interruption of a small blood vessel may result in a speech impediment or difficulty in hearing or an unsteady gait. If a larger blood vessel is involved, the result may be total paralysis of one side of the body. Damage to the right hemisphere of the brain results in disruption of function on the left side of the body, and vice versa. The onset of the stroke may be so sudden and severe that the patient is literally struck down in his tracks. Some patients have early warnings that a stroke may be developing, however.

People who are known to form blood clots in their circulatory system can be given medications to prevent it. Also, current therapy includes medications that can be given to dissolve clots, thereby removing the barrier to blood flow. If blood flow can be reinstated quickly enough, brain tissue may suffer less damage—and less function may be lost.

Strokes can be prevented by effective treatment of high blood pressure and by taking an aspirin tablet every day, for those who can tolerate such medication. The aspirin helps to prevent clot formation, and a number of clinical trials have shown it to be effective in stroke reduction.

Recovery from a stroke varies from one person to the next. Swift treatment followed by effective physical therapy may restore nearly full function of an affected area of the body. Some individuals have experienced severe enough damage that their recovery is minimal and they may be confined to a wheelchair or bed for the remainder of their lives.

[*See also* **Circulatory system; Cognition; Nervous system**]

Brewing

Brewing is the multistage process of making beer and other alcoholic malt beverages. Brewing has taken place around the world for thousands of years, and brewed beverages are staples in the diets of many cultures. Although the main modern ingredients in beer are water, barley, hops, and yeast, people have brewed with products as varied as rice, corn, cassava, pumpkins, sorghum, and millet.

History

Archaeologists have turned up evidence that the Sumerian people in the Middle East were brewing barley grain as long as 8,000 years ago. Ancient Egyptians, Greeks, Romans, Chinese, and Inca also made beer. These early people may have discovered the basic processes of brewing when they observed—and then tasted—what happened to fruit juices or cereal extracts left exposed to the wild yeasts that naturally float in the air.

Over the centuries, breweries sprang up throughout Europe where there was good water for brewing. During the Middle Ages (400–1450), monasteries became the centers for brewing, and the monks originated brewing techniques and created many of the beers still popular today.

A brewmeister and fellow worker inspect the current batch of a local beer in a brewery in the Dominican Republic. *(Reproduced by permission of The Stock Market.)*

Words to Know

Ale: A top-fermented beer that until the latter part of the nineteenth century was not flavored with hops.

Fermentation: Process during which yeast consume the sugars in the wort and release alcohol and carbon dioxide as byproducts.

Hops: Dried flowers of the vine Humulus lupulus, which give beer its characteristic bitter flavor and aroma.

Lager: A traditional Bavarian beer made with bottom-fermenting yeast.

Malt: Barley grain that has germinated, or sprouted, for a short period and is then dried.

Wort: The sugar-water solution made when malted barley is steeped in water and its complex sugars break down into simple sugars.

Yeast: A microorganism of the fungus family that promotes alcoholic fermentation and is also used as a leavening (fermentation) agent in baking.

Bottled beer was introduced by the Joseph Schlitz Brewing Company in Milwaukee, Wisconsin, in 1875. The Gottfried Krueger Brewing Company released the first canned beer in America in 1935.

Brewing process

The basic steps to brewing beer are malting, mashing, boiling, fermentation, aging, and finishing. During malting, barley grains are soaked in water until they begin to germinate, or sprout. The brewer then removes the grains and quickly dries them in a kiln. The dried barley grains are called malted barley or just plain malt. During the mashing phase, the brewer mixes the dried malt with water and heats the mixture until the starchy components in the malt are converted and released into the mixture as simple sugars. The malt is then removed from the mixture, leaving an amber liquid called wort (pronounced wert).

The wort is then heated to a boil and maintained at that temperature for a period of time. During boiling, the brewer adds hops, dried blossoms from the hop plant, which give beer its characteristic bitter flavor and aroma. After the wort is cooled, yeast is added to begin the fermentation stage. These organisms consume the simple sugars in the wort, giving off alcohol and carbon dioxide in the process. The brew is then stored in tanks for several weeks or months while it ages and its flavor develops. To finish the beer, the brewer clarifies the liquid by filtering out the yeast, then packages it in kegs, bottles, or cans.

Types of beer

Beer is usually categorized into two types: ale and lager. Ale is made with a variety of yeast that rise to the top of the fermentation tank and that produce a higher alcohol content than lagers. Ales range from fruity-tasting pale ales to dark and roasty stouts. Lager (from the German word meaning to store) originated in the Bavarian region of Germany. Lager, the most popular beer style in the United States, is made with bottom-fermenting yeast. Lager styles include pilsner (a golden beer with a distinctive hop flavor) and bock (a dark, strong, malty beer).

[*See also* **Fermentation; Yeast**]

Bridges

Bridges are structures that provide a means of crossing natural barriers, such as rivers, lakes, or gorges. Bridges are designed to carry railroad cars, motor vehicles, or pedestrians. Bridges also support pipes, troughs, or other conduits that transport materials, such as an oil pipeline or a water aqueduct.

Humans have been constructing bridges since ancient times. The earliest bridges were probably nothing more than felled trees used to cross rivers or ditches. As civilization advanced, artisans discovered ways to use stone, rock, mortar, and other natural materials to construct longer and stronger bridges. Finally, as physicists and engineers began to develop the principles underlying bridge construction, they incorporated other materials such as iron, steel, and aluminum into the bridges they built. There are four major types of bridges: beam, cantilever, arch, and suspension.

Forces acting on a bridge

Three kinds of forces operate on any bridge: the dead load, the live load, and the dynamic load. Dead load refers to the weight of the bridge itself. Like any other structure, a bridge has a tendency to collapse simply because of the gravitational forces acting on the materials of which the bridge is made. Live load refers to traffic that moves across the bridge as well as normal environmental factors such as changes in temperature,

Words to Know

Abutment: Heavy supporting structures usually attached to bedrock and supporting bridge piers.

Bedrock: Portion of Earth's mantle made of solid rock on which permanent structures can be built.

Dead load: The force exerted by a bridge as a result of its own weight.

Dynamic load: The force exerted on a bridge as a result of unusual environmental factors, such as earthquakes or strong gusts of wind.

Live load: The force exerted on a bridge as a result of the traffic moving across the bridge.

Piers: Vertical columns, usually made of reinforced concrete or some other strong material, on which bridges rest.

Suspenders: Ropes or steel wires from which the roadway of a bridge is suspended.

Truss: A structure that consists of a number of triangles joined to each other.

precipitation, and winds. Dynamic load refers to environmental factors that go beyond normal weather conditions, factors such as sudden gusts of wind and earthquakes. All three factors must be taken into consideration in the design of a bridge.

Beam bridges

The simplest type of bridge consists of a single piece of material that stretches from one side of a barrier to the other side. That piece of material—called a beam or girder—rests directly on the ground on each side or is supported on heavy foundations known as piers. The length of a beam bridge is limited by the weight of the beam itself plus the weight of the traffic it carries. Longer beam bridges can be constructed by joining a number of beams to each other in parallel sections.

Cantilever bridges

A cantilever bridge is a variation of the simple beam bridge. A cantilever is a long arm that is anchored at one end and is free to move at

Caisson

To build bridge piers, workers need a water-free environment to excavate or dig the foundations. This is achieved by using a caisson, a hollow, water-tight structure made of concrete, steel, or other material that can be sunk into the ground. When building a bridge over a river, workers sink a caisson filled with compressed air into the river until it reaches the river bottom. The workers then go into the caisson and dig out soil from the riverbed until they come to bedrock. The caisson, which has sharp bottom edges, continually moves downward during the digging until it comes to rest on bedrock. Concrete is then poured into the caisson to form the lowest section of the new bridge pier.

the opposite end. A diving board is an example of a cantilever. When anchored firmly, a cantilever is a very strong structure. It consists of three parts: the outer beams, the cantilevers, and the central beam. The on-shore edge of the outer beam is attached to the ground itself or to a pier (usually a vertical column of reinforced concrete) that is sunk into the ground. The opposite edge of the outer beam is attached to a second pier, sunk into the ground at some distance from the shore. Also attached to the off-shore pier is one end of a cantilever. The free end of the cantilever extends outward into the middle of the gap between the shores. The cantilevers on either side of the gap are then joined by the central beam.

Trusses. The strength of a cantilever bridge (or any bridge) can be increased by the use of trusses. A truss is structure that consists of a number of triangles joined to each other. The triangle is an important component of many kinds of structures because it is the only geometric figure that cannot be pulled or pushed out of shape without changing the length of one of its sides. The cantilever beam, end beams, and joining beams in a cantilever bridge are often strengthened by adding trusses to them. The trusses act somewhat like an extra panel of iron or steel, adding strength to the bridge with relatively little additional weight. The open structure of a truss also allows the wind to blow through them, preventing additional stress on the bridge from this force.

Arch bridges

The main supporting structure in an arch bridge is one or more curved elements. The dead and live forces that act on the arch bridge are transmitted along the curved line of the arch into abutments or supporting structures at either end. These abutments are sunk deep into the ground, into bedrock if at all possible. They are, therefore, essentially immovable and able to withstand very large forces exerted on the bridge itself. This structure is so stable that piers are generally unnecessary in an arch bridge.

The roadway of an arch bridge can be placed anywhere with relationship to the arch: on top of it, beneath it, or somewhere within the arch. The roadway is attached to the arch by vertical posts (ribs and columns) if the roadway is above the arch, by ropes or cables (suspenders) if the roadway is below the arch, and by some combination of the two if the roadway is somewhere within the arch.

Suspension bridges

In a suspension bridge, thick wire cables run across the top of at least two towers and are anchored to the shorelines within heavy abut-

A bowstring arch bridge in Arizona. The roadway is supported from the arch by suspenders. *(Reproduced by permission of JLM Visuals.)*

ments. In some cases, the roadway is supported directly by suspenders from the cables. In other cases, the suspenders are attached to a truss, on top of which the roadway is laid. In either case, the dead and light loads of the bridge are transmitted to the cables which, in turn, exert stress on the abutments. That stress is counteracted by attaching the abutments to bedrock.

The towers in a suspension bridge typically rest on massive foundations sunk deep into the riverbed or seabed beneath the bridge itself. The wire cables that carry the weight of the bridge and its traffic are made of parallel strands of steel wire woven together to make a single cable. Such cables typically range in diameter from about 15 inches (38 centimeters) to as much as 36 inches (91 centimeters).

Movable bridges

Traditionally, three kinds of movable bridges have been constructed over waterways to allow the passage of boat traffic. In a swing bridge, the roadway rotates around a central span, a large, heavy pier sunk into the river bottom. In a bascule bridge, the roadway is raised like an ancient drawbridge. It can be lifted either at one end or split in two halves in the middle, each half rising in the opposite direction. In a vertical-lift bridge, the whole central portion of the bridge is raised straight up by means of steel ropes.

Brown dwarf

Brown dwarfs—if they indeed exist—are celestial objects composed of dust and gas that failed to evolve into stars. To be a star, a ball of hydrogen must be large enough so that the pressure and heat at its core produce nuclear fusion, the process that makes stars bright and hot. Brown dwarfs, so named by American astronomer Jill Tarter in 1975, range in mass between the most massive planets and the least massive stars, about 0.002 to 0.08 times the mass of the Sun.

Roughly 90 percent of the material in the universe is unaccounted for. Since it cannot be seen, this substance is called dark matter. The existence of dark matter is confirmed by the fact that its mass affects the orbits of objects near the visible edge of galaxies and of galaxies within clusters of galaxies. If brown dwarfs really are as common as astronomers think, their total mass could account for the mass of dark matter, one of modern astronomy's major mysteries.

Words to Know

Cluster of galaxies: A group of galaxies that is bound together by gravity.

Cluster of stars: A group of stars that is bound together by gravity and in which all members formed at essentially the same time.

Dark matter: Unseen matter that has a gravitational effect on the motions of galaxies within clusters of galaxies.

Infrared: Wavelengths slightly longer than visible light, often used in astronomy to study cool objects.

Mass: An object's quantity of matter as shown by its gravitational pull on another object.

Nuclear fusion: Nuclear reactions that fuse two or more smaller atoms into a larger one, releasing huge amounts of energy in the process.

Because brown dwarfs are so cool, small, and faint, they cannot be observed through ordinary telescopes. Beginning in the 1930s, astronomers have suggested their existence using various techniques. One method is to look for a bouncing movement in the path of a star across the sky. Astronomers believe this erratic motion is caused by the gravitational pull of a low-mass companion—such as a brown dwarf—orbiting that star. Another method is to search the sky using infrared telescopes. Some astronomers believe brown dwarfs may emit enough infrared energy to be detected.

A third method astronomers use to locate a suspected brown dwarf is to observe the amount of the element lithium in its spectrum to see if hydrogen fusion reactions are occurring in its core. Lithium is destroyed in the hydrogen fusion reactions of mature stars, but is still present in infant low-mass stars and brown dwarfs. In June 1995, three astronomers reported they found lithium in the spectrum of a suspected brown dwarf called PPL 15 that is located in the cluster of stars known as the seven sisters of the Pleiades (pronounced PLEE-a-dees). Since the stars in the Pleiades cluster are old, the astronomers asserted that PPL 15 is a brown dwarf rather than a low-mass star.

[*See also* **Infrared astronomy; Star**]

Buoyancy

Buoyancy is the tendency of an object to float in a fluid, such as air or water. The principle of buoyancy was first discovered by Greek mathematician Archimedes (c. 287–212 B.C.) and is therefore often called Archimedes' Principle. Legend has it that Archimedes was working on a problem given to him by the king of ancient Syracuse, Hieron II. The king had paid a goldsmith to make him a new crown but suspected that some metal other than gold had been used in the crown. He asked Archimedes to find out if his suspicions were correct—but without destroying the crown.

One day, perhaps while pondering the problem, Archimedes stepped into his bath and noticed the overflow of water. He suddenly realized that the volume of water that had flowed out of the bath had to be equal to the volume of his own body that was immersed. His problem was solved. Later historians claim that Archimedes was so excited with his discovery that he ran naked through the streets of Syracuse shouting "Eureka!" ("I've found it!").

Archimedes began performing experiments with objects in liquids. He immersed the crown and measured the amount of water it displaced. Then he immersed an equal weight of gold in the water. If the crown were pure gold, it would displace the same amount of water. But the crown displaced more water than the gold, indicating it was made of a mixture of gold and silver, which is a bulkier substance. The king had the deceitful goldsmith executed.

What causes buoyancy?

Suppose that you put a block of wood into a container of water. Two competing forces are at work. One force is the downward pressure of the wood on the water. That force is caused by the mass of the wood. The second force is the upward pressure of the water on the block. If the downward pressure of the wood is greater than the upward pressure of the water, the wood sinks. Otherwise, it floats.

The formal statement of Archimedes' Principle is this: the buoyant force acting on an object placed in a fluid is equal to the weight of the fluid displaced by the object. Suppose that the wood block in the example above has a volume of 1 cubic foot. A cubic foot of water weighs about 62.4 pounds. Therefore, water pushes upward on the wooden block with a force of 62.4 pounds.

But a cubic foot of most kinds of wood weighs about 40 to 50 pounds. So water pushes upward more strongly than the wood pushes downward, and the wood floats. If you substituted a block of lead for the wood, the result would be different. A cubic foot of lead weighs about 1,200 pounds. The block of lead pushes downward much more strongly than the water it displaces and it, therefore, will sink.

Practical applications

The principle of buoyancy is used in many forms of transportation. Some sailing ships are made out of materials that are less dense than water and that would, therefore, float under any circumstances. But how can a ship made out of steel float? Steel has a density of 487 pounds per cubic foot, so it would be expected to sink if placed into water.

The reason steel ships float is that they are not constructed of solid pieces of steel. Instead, they consist of hollow shells made of steel. Inside the shell is air, which is much less dense than water or steel. A hollow block of steel and air has an overall density that is less than that of water; therefore, it can float.

Airships such as balloons and dirigibles operate on the same principle. Some material (such as plastic or cloth) is filled with a gas that is less dense than air. The gases most commonly used are hydrogen and helium. The total package of balloon plus gas weighs less than the air it displaces and is, therefore, pushed upward by the air.

[*See also* **Balloon; Density; Pressure**]

Opposite Page: *Termometro Lento,* a buoyancy thermometer, used for determining the concentration of solutions. It works under the principle that the cooler the temperature of the liquid, the denser the solution and the greater its buoyancy. *(Reproduced by permission of Field Mark Publications.)*

Burn

A burn is damage to an area of the body caused by excessive heat, chemicals, electricity, or the Sun. Burns caused by heat can be the result of contact with fire, scalding water, or hot surfaces. Contact with high-voltage power lines can result in electrical burns. A burn can occur on the inside or outside of the body and can range from minor to life threatening.

The severity of a burn is determined by how deeply the injury has penetrated the skin and underlying tissue. If the outer and inner layers of the skin are damaged, a burn is termed partial thickness. If the underlying tissue is also damaged, the burn is referred to as full thickness. However, the depth of a burn is more commonly classified by a system of degrees. The size of a burn is also considered in determining its severity. According to the percentage of the body surface affected, burns are classified as minor, moderate, and major.

The skin

The skin is the largest organ of the body. It covers the outer surface and prevents bacteria, dirt, and other foreign materials from entering the body and causing infection. The skin also aids in regulating body tem-

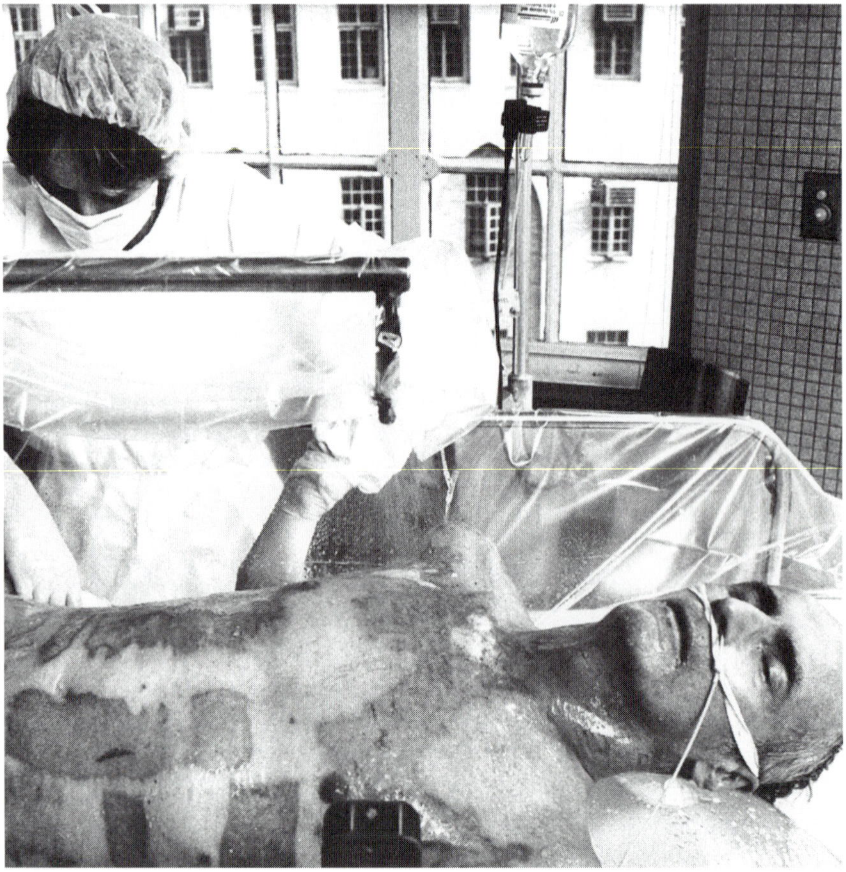

A burn sufferer undergoes debridement (the removal of dead skin). The patterns on his chest are from skin grafts. *(Reproduced by permission of Phototake.)*

Words to Know

Dermis: The layer of skin lying beneath the outer skin.

Epidermis: The outer layer of the skin.

Skin graft: The transfer of skin or skin cells from an uninjured area of the body to a burned area in order to replace damaged skin and promote growth of new skin. A skin graft may be "harvested" from the actual burn victim or may be acquired from a donor.

perature through the process of sweating. Sweat regulates the amount of heat lost from the surface of the body, thus maintaining an even body temperature. In addition, the skin protects against the loss of body fluids that lie beneath it and bathe the tissues.

First-degree burn

A first-degree burn is restricted to the outer layer of the skin, or epidermis, causing redness and pain. A mild sunburn or an injury caused by briefly touching a hot pan are examples of first-degree burns.

Second-degree burn

A second-degree burn is marked by the appearance of blisters on the skin. The burn damage extends through the epidermis to the underlying inner layer, or dermis. The formation of blisters indicates the loss of fluid from cells, and the skin is mildly to moderately swollen and painful. Second-degree burns can be caused by exposure of the skin to sunlight or by contact with a hot object or scalding water.

Third-degree burn

A third-degree burn causes damage to the epidermis, dermis, and the underlying tissue. There are no blisters, but blood vessels in which blood clots have formed are noticeable.

Fourth-degree burn

A fourth-degree burn is one that penetrates the tissue and extends to the underlying muscle and bone. It is the most serious type of burn and is often fatal. Such a burn can result in permanent disfigurement and loss of function of the burned area. Treatment includes replacing lost body fluids, removing destroyed tissue and charred skin, preventing infection, and grafting skin to replace the burned skin. Physical therapy is usually necessary to restore the function of the affected area. A fourth-degree burn results most often from the direct exposure of skin to open flame.

Chemical burns

Chemical burns can occur from ingesting chemicals or having them come in contact with the skin, eyes, or mucous membranes. Some 25,000 industrial chemicals can produce chemical burns. Certain household chemicals, such as drain openers, are also responsible for serious burns.

Butterflies

Butterflies are popular, well-known insects with large, colorful wings covered with tiny scales. Together with moths, butterflies make up the order Lepidoptera, which contains over 150,000 species or kinds. Scientists estimate that about 15,000 butterfly species exist worldwide. During its life cycle, a butterfly undergoes a complete metamorphosis (pronounced met-uh-MORE-fuh-siss) during which it changes from a leaf-eating caterpillar to a nectar-sipping butterfly.

A Monarch butterfly. *(Reproduced by permission of Field Mark Publications.)*

Beauty with a purpose

Butterflies are one of our most favorite insects. The fact that they do not sting or bite, are brightly colored, and do not become pests in people's homes has something to do with why most people enjoy seeing them fly around and would seldom think of killing one (as one might do with other insects thought of only as "bugs.") Most people think that butterflies simply make the world a prettier place. However, like other life forms in the world, they have a place and serve a purpose. For the plant world, butterflies pollinate or carry pollen from plant to plant, helping fruits, vegetables, and flowers to produce new

Words to Know

Chrysalis: A soft casing, shell, or cocoon protecting the dormant pupa of insects during metamorphosis.

Metamorphosis: A complete change of form, structure, or function in the process of development.

Pupa: An insect in the nonfeeding stage during which the larva develops into the adult.

seeds. From the animal point of view, butterflies are near the bottom of the food chain and provide food (especially in their caterpillar stage) for birds, mammals, and other insects.

It is thought that the word butterfly may have originated in England when people started calling the yellow Brimstone or the English sulfur a "butter-colored fly" because the pretty insect reminded them of the color of butter. Eventually it was shortened to "butterfly." The scientific name of its order, Lepidoptera, means "scaly wings" in Latin. This is a correct description since their wings and their bodies are covered with tiny scales. Butterflies and moths are the only insects that have scales. Moths and butterflies are mainly different in their appearance and activities. Moths fly mostly at night and usually have a dull color. Butterflies are active during the day and are brightly colored. Their bodies are thin and hairless, while most moths have plump and furry bodies.

Butterflies are everywhere

Butterflies are found nearly everywhere in the world except Antarctica. They have lived on Earth for at least 150 million years and range in size from the Western Pygmy Blue, which is smaller than a dime and found in North America, to the Queen Alexandra's Birdwing of Papua New Guinea, which has a wingspan of up to 11 inches (28 centimeters). Although some tropical butterflies can live up to one year, the average life span of a butterfly is at most two months. Butterflies display every color of the rainbow in their wings, and no two butterflies are exactly alike. This coloring serves many purposes: from attracting a mate, to blending in with its surroundings, to warning its enemies that it is poisonous and should be avoided.

Unique life cycle

One of the most interesting things about butterflies is their unique life cycle. When a butterfly changes from a slow-moving, fat caterpillar to a colorfully winged, beautiful flying insect, one of nature's most magical events occurs. This metamorphosis happens to most insects, but not as dramatically as it does to a butterfly (the word metamorphosis is Greek for "change in form"). There are four stages in a butterfly's metamorphosis. Every butterfly begins life as an egg. After mating, the female lays her eggs (she actually "glues" them) in small clusters on the leaves of a certain plant. Each species selects its own plant, and the eggs of each are different in shape and markings. In many species, the female dies shortly after doing this. When the egg hatches, the larva emerges. Actually a tiny caterpillar eats its way out the egg, and then proceeds to eat the eggshell. This caterpillar is a true eating machine, and it continues to eat the leaves of the plant where its mother laid her eggs. Caterpillars have one goal—to eat as much as possible—and in their short lifetimes they may eat as much as twenty times their own weight. Caterpillars naturally grow quickly with all this eating, and since their skin cannot stretch, it splits and is shed. This is called molting and it happens several times as the caterpillar gets fatter and fatter. It is at this slow-moving stage that many a caterpillar is devoured by a hungry bird. Still, many protect themselves by using their colors to blend in with their environment. Other have sharp spines or prickly hairs on their bodies to deter predators, while still others have circles or spots on their skin that trick their predators into thinking that the caterpillar is really a larger animal than it is.

Caterpillar to chrysalis to butterfly

If the caterpillar survives and reaches its full size, it attaches itself to a stem from which it hangs upside down. It then sheds its skin one last time, and the old skin hardens almost immediately and becomes a tough shell called a chrysalis (pronounced KRIS-uh-liss). The caterpillar has now become a pupa (pronounced PEW-puh) inside a chrysalis, and its body parts are broken down into a thick liquid that will feed special, preprogrammed cells that have lain dormant in the caterpillar. These cell clusters start to form specialized body parts, like wings, legs, and eyes of a new creature. This process goes on for days, weeks, and sometimes months, according to the type of butterfly that will emerge. The final stage occurs when an adult butterfly finally pushes itself out of its chrysalis, looking nothing like the caterpillar that it was. When the butterfly breaks through the now-soft shell, its wings are wet and crumpled and it must rest while it expands its wings and pumps them full of blood. Continued flapping makes them strong, and soon the adult butterfly is ready to fly away and begin this cycle all over again by looking for a mate.

Every adult butterfly is covered with millions of tiny scales that help it to control its body temperature. They also can help it escape from a predator's grip since they rub off easily. It is these scales that give butterflies their beautiful colors. A butterfly's body is made up of three parts: head, thorax, and abdomen. On its head are two long antennae, which it uses as feelers to touch and to smell things. They also have two large compound eyes, which means that each is really thousands of eyes formed together. This allows a butterfly to see in all directions at once. They also have a long hollow tube called a proboscis (pronounced pro-BOSS-siss), which they use like a straw to sip the energy-rich, sugary liquid called nectar produced by flowering plants.

Amazing fliers

A butterfly's wings are its most important part since they enable it to move about for food, shelter, a mate, and all the other things it needs. Its wings are very strong, and they are supported and shaped by a network of veins, just like those in a leaf. Different species have different-shaped wings that make each fly in a different manner. Those with large wings flap and make long glides, while those with wide wings flutter and flit or move with short bursts. Those with long, thin wings fly the fastest, and those with short, triangular wings can zigzag and dart about quickly. No matter how they move about, butterflies are incredible fliers, and some migrate over 3,000 miles (4,800 kilometers) to spend the winter in a warmer place. The well-known Monarch butterfly flies to Mexico from North America before the autumn chill arrives. Those living east of the Rocky Mountains fly the longest, traveling over 2,000 miles (3,200 kilometers) to get to the same spot in Central America. Some scientists believe that they find their way by using the position of the Sun as a compass, while others think that they are able to detect changes in light waves that are filtered through the clouds. However they do it, millions make the journey south every year, and their offspring make their way north again the following spring.

The ancient Greeks are said to have believed that when people die, their souls leave their bodies in the form of a butterfly. Their symbol for the soul was a young girl named Psyche who had butterfly wings. Today, we know that real butterflies are extremely sensitive to changes in their environment. More and more, as their habitats are being destroyed and endangered by pollution, pesticides, and other human activity, butterflies are threatened. Some rare species may have already become extinct.

[*See also* **Insects; Invertebrates**]

C

CAD/CAM

CAD/CAM is an acronym for computer-aided design and computer-aided manufacturing. The use of computer software programs makes it possible to remove much of the tedious manual labor involved in design and manufacturing operations. Prior to CAD/CAM software programs, engineers building complex machines had to prepare thousands of highly

A CAD (computer-aided design) system used for Boeing airplanes. *(Reproduced by permission of Phototake.)*

technical and accurate drawings and charts. When changes had to be made to the plans during the building process, engineers and draftspersons had to make them by hand. This was time-consuming, and often introduced errors to the original plans. When a CAD program is used, the computer automatically evaluates and changes all corresponding documents instantly and without errors.

CAD programs also allow engineers and architects to test new ideas or designs without having to actually build the design. This has been particularly valuable in space programs, where many unknown design variables are involved. Previously, aerospace engineers depended upon trial-and-error testing, a time-consuming and possibly life-threatening process. Computer simulation and testing in the space program reduces time and the possible threat to lives. CAD is also used extensively in the military, civil aeronautics, automotive, and data processing industries.

CAM, commonly operated together with CAD, is a software program that communicates instructions to automated machinery. CAM techniques are especially suited for manufacturing plants, where tasks are repetitive or dangerous for human workers.

[*See also* **Automation; Robotics**]

A modern hand-held calculator (foreground) and its predecessor. *(Reproduced by permission of The Stock Market.)*

Calculator

The calculator is a computing machine. Basic calculators perform basic mathematical functions (addition, subtraction, division, and multiplication). More sophisticated calculators can perform functions of the higher-based mathematical branches of trigonometry and calculus. The odometer, or mileage counter, in an automobile is a counting machine, as is the pocket calculator and the personal computer. They may have different ability levels, but they all tally numbers.

Early calculators

Although the abacus, the first tool of calculation, has existed since ancient times, advanced calculating machines did not appear until the early 1600s. Blaise Pascal, a French scientist and philosopher, developed in 1642 the

Pascaline, a machine capable of adding and subtracting nine-digit numbers. Figures were entered by moving numbered wheels linked to each other by gear, similar to a modern automobile's odometer. In 1671, German philosopher and mathematician Gottfried Wilhelm Leibniz improved Pascal's design, creating a machine that performed multiplication. In 1820, Frenchman Charles X. Thomas devised a machine that added subtraction and division to a Leibniz-type calculator. It was the first mass-produced calculator, and it became a common sight in business offices.

Over the next century, mathematicians and inventors improved upon the designs of previous calculating machines. In 1875, American inventor Frank Stephen Baldwin received the first patent for a calculating machine. Baldwin's machine did all four basic mathematical functions and did not need to be reset after each computation. With the need for more accurate record keeping in the business world, calculating machines that used motors to tally larger and larger numbers and mechanisms to print out results on paper were devised. These mechanical machines remained essentially unchanged until the mid-1960s.

Electronic calculators

The integrated circuit chip—tiny, complex electronic circuits on a single chip of silicon—was invented in 1959 by Texas Instruments and Fairchild (a semiconductor manufacturing company). Although integrated circuits allowed calculators to become much faster and smaller, those early electronic calculators were still just adding machines. In 1970, however, the development of the microprocessor—which incorporated the circuitry of the integrated circuit and the entire central processing unit of a computer onto a single chip—changed the computing industry. The microprocessor made pocket-sized, highly sophisticated calculators possible.

Today, pocket calculators with a wide range of functions are available, including programmable calculators that are in effect miniature computers. Some calculators are powered by solar cells in ordinary room light. More than 50 million portable calculators are sold in the United States each year, many for less than $10.

[*See also* **Computer, digital**]

Calculus

Calculus is a field of mathematics that deals with rates of change and motion. Suppose that one nation fires a rocket carrying a bomb into the atmosphere, aimed at a second nation. The first nation must know exactly what path the rocket will follow if the attack is to be successful. And the second nation must know the same information if it is to protect itself against the attack. In this example, calculus is used by mathematicians in both nations to study the motion of the rocket.

Calculus was originally developed in the late 1600s by two great scientific minds, English physicist Isaac Newton (1642–1727) and German mathematician Gottfried Wilhelm Leibniz (1646–1716). Both scholars presented their ideas at about the same time, so credit for the invention of calculus must go to both. The debate over credit at the time, however, reached intense levels and sparked bad feelings between the two countries involved (Great Britain and Germany). Over the past 300 years, calculus has become an absolutely essential mathematical tool in every field of science, mathematics, and engineering.

To illustrate the basic principles of calculus, imagine that you are studying changes in population in your hometown over the past 100 years. As you graph the data you collected, you can see that population increased for a number of years, then decreased for a period of time before beginning a second increase. One question you might want to ask is what the rate of change in the population was at any given time, such as any given year. For example, was population increasing at the same rate in 1980 that it was in 1890? One way to answer that question is to locate two points on the curve. The rate of change for this part of the graph, then, is how steeply the curve rises between these two points.

Differential and integral calculus

Calculus can be subdivided into two general categories: differential and integral calculus. Differential calculus deals with problems of the type above, in which some mathematical function (such as population change) is known. From the graphical or mathematical representation of that function, the rate of change can be calculated.

The reverse process can also be performed. For example, it may be possible to find the rate of change for some function. From that rate of change, then, it may be possible to determine the original function itself. This field of mathematics is known as integral calculus.

Calendar

A calendar is a system of measuring the passage of time for the purpose of recording historic events and arranging future plans. Units of time are defined by three different types of motion: a day is one rotation of Earth around its axis; a month is one revolution of the Moon around Earth; and a year is one revolution of Earth around the Sun. The year is the most important time unit in most calendars, since the cycle of seasons repeat in a yearly cycle as Earth revolves around the Sun.

Making a yearly calendar, however, is no simple task because these periods of time do not divide evenly into one another. For instance, the Moon completes its orbit around Earth (a lunar month) in 29.5 days. A lunar year (12 lunar months) equals 365 days, 8 hours, and 48 minutes. A solar year (time it takes Earth to complete its orbit around the Sun) is 365.242199 days, or 365 days, 5 hours, 48 minutes, and 46 seconds. The calendar we currently use is adjusted to account for the extra fraction of a day in each year.

Development of the present-day calendar

The official calendar currently used worldwide is the Gregorian calendar. The ancient Egyptians adopted a 365-day calendar sometime between 4000 and 3000 B.C. The first major improvement to that 365-day calendar was made by Roman dictator Julius Caesar (100–44 B.C.) in 46 B.C. With the help of Greek astronomer Sosigenes, Caesar developed a calendar divided into 12 months of 30 and 31 days, with the exception of 29 days in February. In this new Julian calendar (named after Caesar), an extra day, or leap day, was added to every fourth year to account for the 365.25-day solar year.

The Julian calendar, however, was still off by 11 minutes and 14 seconds each year. Over 300 years, this difference added up to just over 3 days. By the mid-1500s, the Julian calendar was another 10 days ahead of Earth's natural yearly cycle.

To adjust this calendar to line up with the seasons, Pope Gregory XIII (1502–1585) introduced another change in 1582. He first ordered that 10 days be cut from the current year, so that October 4, 1582, was followed by October 15, 1582. He then devised a system whereby three days are dropped every four centuries. Under the original Julian calendar, every century year (200, 300, 400, etc.) was a leap year. In the new calendar, named the Gregorian calendar, only those century years divisible by 400 (800, 1200, 1600, etc.) are leap years.

Although not perfect, the Gregorian calendar is accurate to within 0.000301 days (26 seconds) per year. At this rate, it will be off one day by about the year 5000.

The Jewish and Muslim calendars

The Jewish calendar is a lunisolar calendar, a combination of lunar and solar years. The calendar is 12 lunar months long, with an additional month added every few years to keep the calendar in line with the seasons. The months (with corresponding days) are Tishri (30), Marheshvan (29 or 30), Kislev (29 or 30), Tebet (29), Sebat or Shebat (30), Adar (29), Nisan (30), Iyar (29), Sivan (30), Tammuz (29), Ab (30), and Elul (29). Adar II, the extra month, is added periodically after Adar. Calendar years vary from 353 to 355 days; leap years may have 383 to 385 days. Rosh ha-Shanah, the Jewish New Year, is observed on the first day of Tishri.

The Muslim calendar is a strict lunar calendar. Calendar years vary from 354 to 355 days, with the months and seasons having no connection. The months are Muharram (30), Safar (29), First Rabia (30), Second Rabia (29), First Jumada (30), Second Jumada (29), Rajab (30), Shaban (29), Ramadan (30), Shawwal (29), Dhu-I-Kada (30), and Dhu-I-Hijja

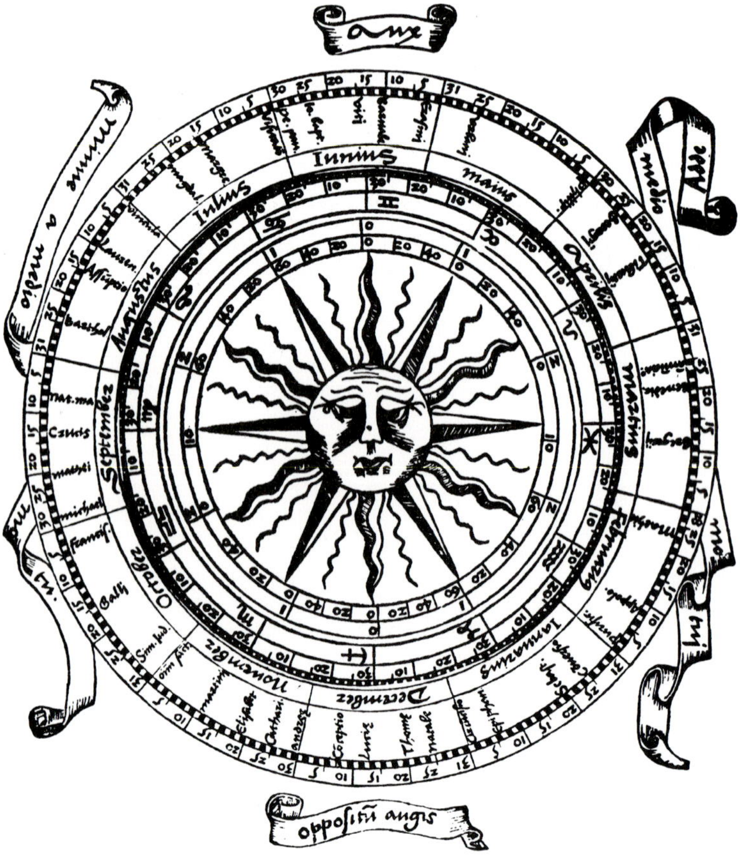

An early Roman calendar. *(Reproduced by permission of Corbis-Bettmann.)*

(29 or 30). The first day of the first year of the Muslim calendar corresponds to July 16, 622, of the Gregorian calendar.

The Gregorian calendar and the third millennium

Technically, the year 2000 on the Gregorian calendar was not the beginning of the third millennium (a 1,000-year span). In actuality, it was the last year of the second millennium. When the Gregorian calendar was adopted, the transition of the years B.C. to A.D.—marking the birth of Jesus of Nazareth—did not include the year 0. The sequence runs 2 B.C., 1 B.C., A.D. 1, A.D. 2, etc. According to this sequence, since there is no year zero, the first year of the first millennium was A.D. 1. Thus, the first day of the third millennium was January 1, 2001.

Possible future calendar reform

Although the Gregorian calendar allows for the oddity of Earth's orbit and for the dates when Earth is closest and farthest from the Sun, the shortness of February introduces slight problems into daily life. For example, a person usually pays the same amount of rent for the 28 days of February as is paid for the 31 days of March. Also, the same date falls on different days of the week in different years. These and other examples have led to several suggestions for calendar reform.

Perhaps the best suggestion for a new calendar is the World Calendar, recommended by the Association for World Calendar Reform. This calendar is divided into four equal quarters that are 91 days (13 weeks) long. Each quarter begins on a Sunday on January 1, April 1, July 1, and October 1. These 4 months are each 31 days long; the remaining 8 months all have 30 days. The last day of the year, a World Holiday (W-Day), comes after December 30 (Saturday) and before January 1 (Sunday) of the new year. W-Day is the 365th day of ordinary years and the 366th day of leap years. The extra dy in leap years would appear as a second World Holiday (Leap year or L-Day) between June 30 (Saturday) and July 1 (Sunday). The Gregorian calendar rules for ordinary, leap, century, and noncentury years would remain unchanged for the foreseeable future.

Calorie

A calorie is a unit of heat measurement in the metric system. It is defined as the amount of heat needed to raise one gram of pure water by 1°C under standard conditions. The term standard conditions refers to atmospheric pressure of one atmosphere and a temperature change from 15.5 to 16.5°C.

A second unit of measurement is the Calorie. The Calorie (with a capital C) is 1,000 times the size of the calorie. The difference between the two is sometimes made clear by calling the calorie the gram-calorie and the Calorie the kilocalorie. The abbreviation for the two units are cal for the gram-calorie and Cal for the kilocalorie. When you read about the number of calories contained in food or the number of calories to include in your diet each day, the term intended is the kilocalorie. It is this unit that is used by nutritionists in talking about the food value of what we eat.

Many people other than health scientists are interested in the caloric content of substances. For example, engineers need to know the heat content (in calories) of various types of fuels. Ecologists are interested in the energy content (expressed in calories) of various organisms in the environment and in how that energy content changes over time.

One calorie in the metric system is equivalent to 3.968 British thermal units, the unit of measurement for heat energy in the British system. A calorie is also equivalent to 4.187 joules, the fundamental unit of heat energy in the *Système International,* or International System of Units, the measurement system used by scientists throughout the world.

[*See also* **Energy; Units and standards**]

Canal

A canal is a human-made waterway or channel that is built for transportation, irrigation, drainage, or water supply. Although canals are among the oldest works of civil engineering, they continue to play a major role in commerce, as they are the cheapest form of inland transportation yet devised.

The earliest canals were built by Middle Eastern civilizations primarily to provide water for drinking and for irrigating crops. The Nahrwan Canal, 185 miles (300 kilometers) long, was built around 2400 B.C. between the Tigris and Euphrates Rivers (in present-day Iraq). Egypt's ancient pharaohs linked the Mediterranean and Red Seas with a canal that the Romans later restored and used for shipping.

The Chinese were perhaps the greatest canal builders of the ancient world, having linked their major rivers with a series of canals dating back to the third century B.C. Their most impressive project was the famous Grand Canal, begun in the sixth century B.C. and completed in the thir-

Words to Know

Contour canal: Usually early canals that followed the meandering natural contours of the landscape.

Inland canal: An artificial waterway or channel that is cut through land to carry water and is used for transportation.

Lock: A compartment in a canal separated from the main stream by watertight gates at each end; as water fills or drains it, boats are raised or lowered from one water level to another.

teen century. Stretching for a total length of 1,000 miles (1,600 kilometers), the Grand Canal is the longest canal in the world.

Transportation

Canal systems for transportation were not widespread in Europe until the seventeenth and eighteenth centuries. The famous Canal du Midi in southern France allowed oceangoing ships to travel 150 miles (240 kilometers) from the Atlantic Ocean to the Mediterranean Sea. The completion of this canal in 1681 spurred the construction of transportation canals in England, Germany, and other European countries. With the onset of the Industrial Revolution in Europe in the eighteenth century, transport by canal soon became essential to the movement of raw materials and manufactured goods throughout Europe.

The first major canal in the United States was the Erie Canal. Completed in 1824, the 364-mile (586-kilometer) canal provided a water route that brought grain from the Great Lakes region to New York and the markets of the East. With the coming of the railroads in the 1830s, the United States quickly abandoned its canals, believing that rail was then the best method for shipping and transportation. Today, however, inland canals play a major transportation role in the United States and the rest of the world, since they are perfectly suited for carrying low-value, high-bulk cargoes over long distances.

Construction and operation

The paths of most canals are affected by variations in the landscape through which they pass. Very early canals followed the lay of the land, simply going around anything in their way. Because of this, they were called contour canals. These canals could not, however, connect two bodies of water that were at different heights. The invention of the lock system in China in the tenth century solved this engineering problem, opening the way for the full development of canals.

In the modern lock system, a segment of a canal is closed off by gates at either end. When a boat enters the lock, the front gate is already closed. The back gate is then closed behind the boat, and the water level

A canal boat in a narrow canal in the Netherlands. *(Reproduced by permission of Photo Researchers, Inc.)*

within the lock is raised or lowered to the level of the water on the outside of the front gate. Valves on the gates control the level of the water.

Sea canals, the great canals that shorten sea routes, are engineering achievements. Three outstanding examples are the Suez Canal linking the Mediterranean Sea and the Gulf of Suez (1869), the Kiel Canal connecting the North and Baltic Seas (1895), and the Panama Canal linking the Atlantic and Pacific Oceans (1914).

[*See also* **Lock**]

Cancer

Cancer is a disease of uncontrolled cell growth caused by exposure to carcinogens (cancer-causing substances), genetic defects, or viruses. Cancer cells can multiply and form a large mass of tissue called a tumor. Some tumors are limited to one location and can be surgically removed. These tumors may cause little harm and are therefore termed benign. Cancer cells of other tumors may spread, or metastasize (muh-TASS-tuh-size), to surrounding tissue or other organs of the body. Such aggressive tumors are termed malignant. Cancer is a word used usually to describe malignant, not benign, tumors. The study of cancer is called oncology.

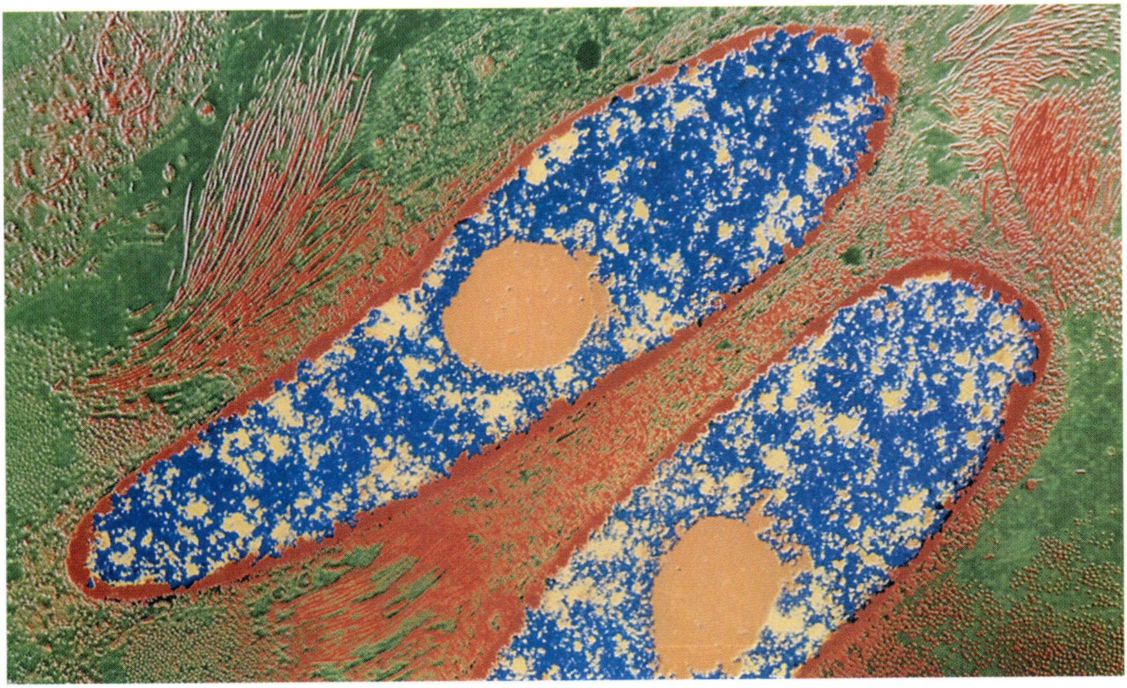

A transmission electron micorgraph of two spindle cell nuclei from a human sarcoma. Sarcomas are cancers of the connective tissue (such as bone, cartilage, tendons, and ligaments). *(Reproduced by permission of Photo Researchers, Inc.)*

Words to Know

Carcinogen: A substance capable of causing cancer.

Chromosome: Organized strands of DNA in the nucleus of a cell.

DNA (deoxyribonucleic acid): The genetic material in the nucleus of cells that contains information for an organism's development.

Gene: A section of a chromosome that carries instructions for the formation, functioning, and transmission of specific traits from one generation to another.

Mutation: A change in the DNA in a cell.

Tumor: A mass of abnormal tissue that can be malignant or benign.

How cancer cells are formed

The transformation of a normal cell into a cancer cell can occur when the genetic material (deoxyribonucleic acid or DNA) of a cell is changed, or mutated. A tumor is the result of multiple gene mutations within a single cell. Years or decades before a tumor forms, a cell can become weakened by various factors, making it more susceptible to later transformation into a cancer cell. Cancer is often a disease of age, with many occurring after age fifty.

Types of cancer

There are more than 200 different types of cancer, and they are named for the organ or tissue in which they begin to grow. Leukemia refers to cancer of white blood cells (also called leukocytes), and lymphoma is cancer of lymphoid tissue (a connective tissue containing white blood cells called lymphocytes). Melanomas are cancers that begin in melanocytes (skin pigment cells). Cancers that originate in epithelial tissue (cellular tissue that lines cavities such as the stomach or lung) are called carcinomas. Those that begin in connective tissue (such as bone and cartilage) or muscle are called sarcomas.

Causes of cancer

One of the most carcinogenic substances known is tobacco smoke. It is the major cause of lung cancer, which is the leading cause of cancer deaths in both men and women in the United States. In the year 2000, almost 160,000 people in the United States died from lung cancer. In comparison, cancer of the colon and rectum caused over 56,000 deaths that same year. Breast cancer claimed over 41,000 lives, while prostate cancer claimed almost 32,000.

Other carcinogens include certain chemicals, the Sun's ultraviolet light, and radiation. Some viruses can cause cancer by altering the DNA of a host cell and converting the cell's normal genes into cancer-causing genes, or oncogenes. Genetic factors—such as chromosomal

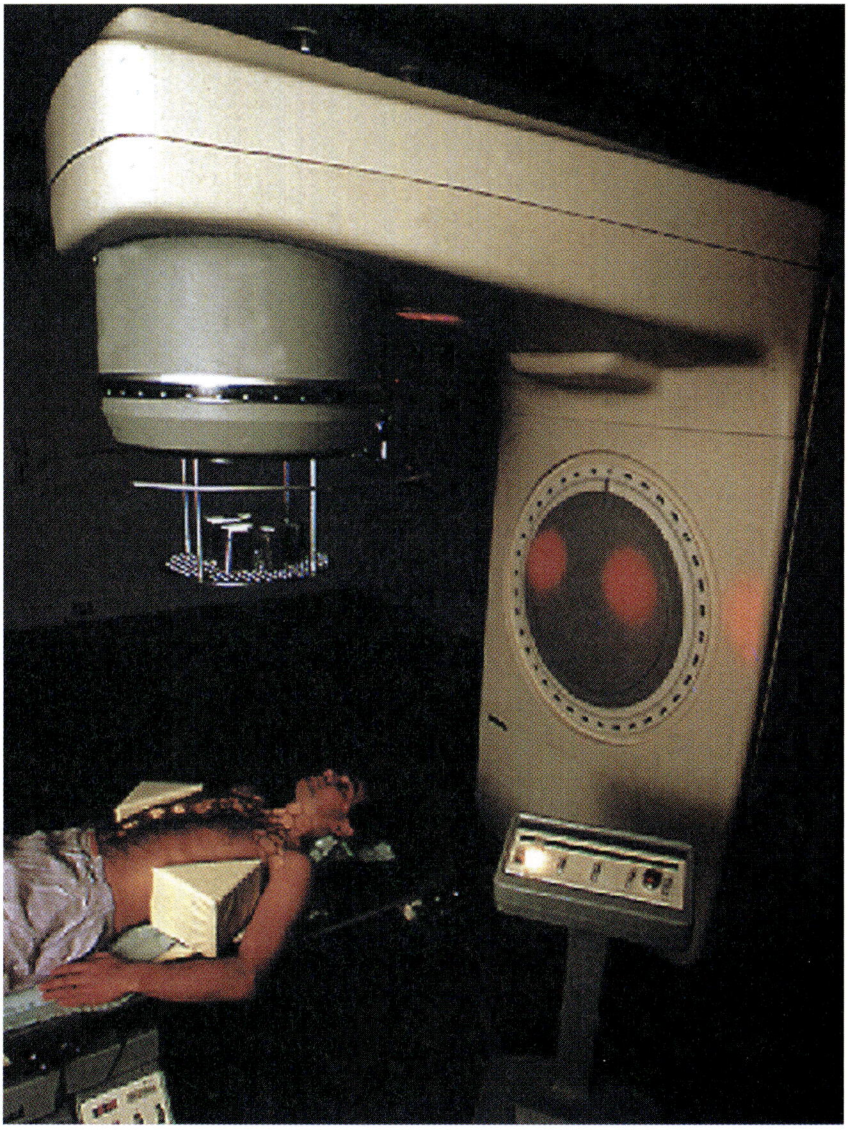

Man undergoing radiation treatment for Hodgkin's disease, a cancer of the lymphatic tissue. The areas to be irradiated appear as illuminated circles on the patient's body. *(Reproduced by permission of Photo Researchers, Inc.)*

abnormalities or the inheritance of faulty genes from a parent—can make some people more likely to develop certain cancers. For instance, people with Down syndrome, a chromosomal abnormality, are susceptible to leukemia.

Treatment

Cancer treatment consists of surgery to remove tumors and radiation to slow tumor growth. Chemotherapy, or drug therapy, is often used to treat cancers that have spread to other parts of the body. Another approach to treatment is to boost the immune system with immune-enhancing drugs or antibodies that can recognize and destroy abnormal cells. A new type of anticancer drug that homes in on cancer cells—leaving healthy ones alone—was introduced in the spring of 2001. The drug, Gleevec, works strikingly well against chronic myelogenous leukemia (one of the four main types of blood cancer) and gastrointestinal stromal tumor (GIST; a rare stomach tumor). Scientists believe the future of cancer treatment lies in the development of a wider range of anticancer drugs and in genetic engineering, whereby healthy genes would be manufactured to replace mutated DNA in transformed cells.

Prevention

Many cancers are preventable. It is estimated that avoidance of tobacco, overexposure to the Sun, high-fat diets, excessive alcohol, unsafe sex, and other known carcinogens could prevent more than 80 percent of all cancer cases. In addition, yearly testing can detect certain cancers and make early treatment possible, providing a better chance of survival.

[*See also* **Carcinogen; Nucleic acid**]

Canines

Biologists classify canines as members of the carnivore (meat-eating) family Canidae. That family, made up of 30 to 35 species, includes wolves, foxes, coyotes, the dingo, jackals, and a number of species of wild dog. The family also includes the domesticated dog, which is believed to have descended from the wolf.

Canines originated 38 million to 54 million years ago in North America, from where they spread throughout the world. Canines range in size from the fennec fox, which is about 16 inches (40 centimeters) long, including the tail, and which weighs 3 pounds (1.4 kilogram), to the gray wolf, which is more than 6 feet (2 meters) in length and which weighs about 175 pounds (80 kilograms). Canine skulls have a long muzzle, well-developed jaws, and a characteristic dental formula of 42 teeth.

The social behavior of canines varies from solitary habits to highly organized packs that exhibit considerable cooperation. Most canines live in packs, which offers several benefits including defense of group territory, the care of the young, and the ability to catch large prey species.

Wolves

Wolves are found in North America, Europe, and Asia. The gray or timber wolf is the largest member of the dog family and a widely distributed species. It lives in a variety of habitats, including forests, plains, mountains, tundra, and deserts. The red wolf is found only in southeast-

The endangered red wolf. *(Reproduced by permission of Photo Researchers, Inc.)*

ern Texas and southern Louisiana. The red wolf is smaller than the gray wolf, and there is some evidence that it may be even a hybrid cross between the gray wolf and coyote.

The gray wolf lives in packs and is a territorial species. Territories are scent-marked (that is, an animal will leave its personal odor to mark the boundaries of its territory) and may range from 50 to 5,000 square miles (130 to 13,000 square kilometers). Pack size is usually small, about eight members, consisting of a mature male and female and their offspring and relatives. When the pack size becomes large, a system of dominance hierarchy is established. The dominant male leader of the pack is called the alpha male and the dominant female is called the alpha female. Hierarchy is acknowledged among pack members through submissive facial expressions and body postures.

Only the dominant male and female breed. Gestation (pregnancy) is about two months and the average litter size is from four to seven pups, who are born blind. The young reach physical maturity within the year. They do not become sexually mature until the end of their second year. The nonbreeding members of a pack help protect and feed the young. Wolves prey on deer, moose, elk, and beavers.

Besides scent-marking, wolves communicate by howling. Scientists believe that howling informs pack members of each other's position when they are separated from each other and warns off other packs from the territory as well. During spring and summer, the wolf pack remains within its territory. Wolf pups are raised during this period. During autumn and winter, the wolf pack travels widely, often following the migration of species on whom they prey.

Foxes

There are 21 species of fox. Foxes range in size from the 3-pound (1.4-kilogram) fennec fox to the 20-pound (9 kilogram) red fox. The gray fox and arctic (or white) fox are highly valued for their pelts. The so-called silver and blue fox are actually color phases of the arctic fox. Other fox species include the kit fox, the swift fox, crab-eating South American fox, and the sand fox.

Foxes have a pointed muzzle, large ears, a slender skull, and a long bushy tail. They are territorial and, like wolves, scent-mark their territories. Foxes also communicate by vocalizations such as yapping, howling, barking, whimpering, and screaming. Foxes use stealth, and dash-and-grab hunting techniques to catch their prey. They are generally solitary hunters, and most species feed on rabbits, rodents, birds, beetles, grasshoppers, and earthworms. The bat-eared fox of Africa eats primarily insects, although it also includes fruit and small animals in its diet. Foxes mate in winter, with the female producing an annual litter of one to six pups after a gestation period of fifty to sixty days.

Foxes are heavily hunted for their pelts and to prevent the spread of the viral disease rabies. Some attempts to develop oral vaccination for rabies have been successfully used in Switzerland and Canada.

Other members of the Canidae family

Coyotes, jackals, dingoes, and a number species of wild dog comprise the rest of the canine family. The distribution of coyotes ranges from Alaska to Central America. Their populations have flourished as wolves have been eliminated. They have a tendency to interbreed with wolves and with domestic dogs. Coyotes prey on small animals, but will eat whatever is available to them, including carrion (dead meat), insects, and fruit.

Dingoes are wild dogs that originally lived in Australia. *(Reproduced by permission of Photo Researchers, Inc.)*

Coyotes reach maturity during their first year and produce a litter of six pups. While the basic social unit of coyotes is the breeding pair, some coyotes form packs similar to wolf packs and scent-mark a territory. In the United States, coyotes have been responsible for considerable losses of sheep.

Four species of jackals occupy ecological niches normally filled by wolves in warmer parts of the world. They are found throughout Africa, southeastern Europe, and southern Asia as far east as Burma. The four species are the golden jackal, the simian jackal, the black-backed jackal, and the sidestriped jackal. The golden jackal is the most widely distributed of the four species.

The golden jackal prefers arid grasslands, the silverbacked jackal prefers brush woodlands, the simian jackal prefers the high mountains of Ethiopia, and the sidestriped jackal prefers moist woodlands. Jackals have a varied diet of fruit, reptiles, birds, and small mammals.

Jackals are unusually stable in their breeding relationships, forming long-lasting partnerships. They also engage in cooperative hunting. Jackals are territorial and engage in scent-marking, usually as a unified male and female pair. They also participate in raising their young together and tend to remain monogamous (staying with the same mate). Jackals communicate by howling, barking, and yelping. In Ethiopia, the simian jackal is an endangered species because it has been killed for its fur.

Other wild canines include the Indian wild dog dhole of southeast Asia and China; the maned wolf; the bush dog of Central America and northern parts of South America; the dingo of Australia; and the raccoon dog of east Asia, Siberia, Manchuria, China, Japan, and the northern Indochinese peninsula. In Africa, the cape hunting dog has a distinctive black, yellow, and white coat and large ears and hunts in packs that can overpower any game species.

Domestic dogs

Kennel societies in the United States recognize 130 breeds of domesticated dog, while kennel societies in Great Britain recognize 170 breeds, and the Federation Cynologique Internationale (representing 65 countries) recognizes 335 breeds. Domestic dogs range in size from about 4 pounds (2 kilograms) to more than 200 pounds (100 kilograms). Some breeds, such as the dachshund, have very short legs, while others, such as the greyhound, have very long legs.

For dog owners, these animals serve a number of different purposes. Pet dogs provide companionship and protection, while others herd sheep or cattle or work as sled dogs. Police use dogs to sniff out illegal drugs and to help apprehend criminals. Dogs are also used for hunting and for racing. Guide dogs help blind people find their way around.

Female dogs are first able to reproduce, at the age of anywhere from seven to eighteen months. Gestation lasts about two months, and the size of a litter is from three to six puppies. Like other canines, puppies are born unable to see. They develop that sense in about twenty-one days. At about the age of two months, puppies are less dependent on their mothers and begin to relate more to other dogs or people. Typical vocalizations of domestic dogs include barking and yelping.

History of domestication. About 5,000 years ago, human civilization changed from a hunting-gathering society to a farming culture, and the domestication of the dog began. Scientists believe that all breeds of domestic dogs, whether small or large, long haired or short haired, are descended from a wolflike animal over several millions of years. Domestic dog breeds have been produced through selective breeding.

Unlike the wolf, the dog is treasured by humans, as indicated by its nickname as "man's best friend." Dog stories abound in children's literature, from Lassie to Rin Tin Tin, and politicians, like Presidents Franklin D. Roosevelt and Richard Nixon, used their pet dogs to enhance their images.

Endangered canines

Hunting and the destruction of habitat have endangered some species of canines. Their reputation as predators has added to efforts to eradicate them from areas where livestock are raised or where they live too close to urban populations. Some rare wild dogs, such as the jackal of Ethiopia, are few in number. The maned wolf of Argentina and Brazil has a population of no more than 1,000 to 2,000 members. A successful effort has been made to reintroduce the gray wolf into national parks in the United States, but this effort is opposed by many people, particularly local ranchers.

Carbohydrate

Carbohydrates are compounds composed of carbon, hydrogen, and oxygen. The carbohydrate group includes sugars, starches, and cellulose. Sugars and starches provide organisms with energy for cell functions, and cellulose is a fibrous substance making up plant cell walls.

The role of plants

Carbohydrates are produced by green plants through a process known as photosynthesis. In photosynthesis, a pigment called chlorophyll (pronounced KLOR-uh-fill) in the leaves of plants absorbs light energy from the Sun. Plants use this light energy to convert water and carbon dioxide from the environment into glucose and oxygen. Some glucose is used to form the more complex carbohydrate cellulose, the main structural component of plant cell walls. Some is used to provide immediate energy to plant cells. The rest is changed to a different chemical form, usually starch, and stored in seeds, roots, or fruits for later use.

The carbohydrates produced by plants are an important source of energy for animals. When animals eat plants, energy stored in carbohydrates is released in the process of respiration, a chemical reaction between glucose and oxygen to produce energy (for cell work), carbon dioxide, and water. Glucose is also used by animal cells in the production of other substances needed for growth.

Types of carbohydrates

Carbohydrates are usually divided into three categories: monosaccharides, having one sugar unit; disaccharides, having two sugar units; and polysaccharides, having many sugar units. The arrangement of atoms in different carbohydrate molecules gives each its specific properties.

Among the most important monosaccharides are glucose (found in plants and animals), fructose (a fruit sugar), and galactose (formed from the milk sugar lactose). Glucose is especially important in vital life processes. Each of these simple sugars is made up of the same number and types of atoms, arranged differently. All three are readily absorbed into the bloodstream from the digestive tract.

Disaccharides are carbohydrates composed of two subunits of simple sugars. They include sucrose (ordinary table sugar), maltose (formed from starch), and lactose (found in milk and the only sugar not occurring in plants). Enzymes in the digestive tract of humans and animals split disaccharides into the more easily absorbed monosaccharides. For example, sucrose is split into the simple sugars glucose and fructose during digestion.

Polysaccharides are the most abundant carbohydrates. A polysaccharide molecule can contain thousands of glucose units. These highly complex carbohydrates include starches, cellulose, and glycogen. Starch is the usual form in which carbohydrates are stored as energy by plants. Plants can split starch into its simpler glucose units for use as energy when needed.

Cellulose is known as a structural carbohydrate because of the fibers formed by its molecules. It is the major component of plant cell walls and comprises over half of the carbon found in plants. Humans and most animals cannot digest cellulose for food but use it as a fiber (often termed roughage) to help in the elimination of waste by the intestine. Some grazing animals such as cows and sheep have microorganisms in their digestive tracts that partially digest cellulose, allowing these animals to use cellulose as food.

Glycogen is the form in which glucose is stored in the liver and muscles of animals for energy needs. Muscle glycogen is used primarily to fuel muscle contractions (such as during exercise). Liver glycogen is used to restore glucose to the blood when the sugar level is low (such as between meals).

[*See also* **Photosynthesis**]

Carbon cycle

The carbon cycle is the process in which carbon atoms are recycled over and over again on Earth. Carbon recycling takes place within Earth's biosphere and between living things and the nonliving environment. Since a continual supply of carbon is essential for all living organisms, the carbon cycle is the name given to the different processes that move carbon from one to another. The complete cycle is made up of "sources" that put carbon back into the environment and "sinks" that absorb and store carbon.

Recycling carbon

Earth's biosphere can be thought of as a sealed container into which nothing new is ever added except the energy from the Sun. Since new matter can never be created, it is essential that living things be able to reuse the existing matter again and again. For the world to work as it does, everything has to be constantly recycled. The carbon cycle is just one of several recycling processes, but it may be the most important process since carbon is known to be a basic building block of life. As the foundation atop which a huge family of chemical substances called organic substances are formed, carbon is the basis of carbohydrates, proteins, lipids, and nucleic acids—all of which form the basis of life on Earth.

Since all living things contain the element carbon, it is one of the most abundant elements on Earth. The total amount of carbon on Earth, whether we are able to measure it accurately or not, always remains the

Words to Know

Biosphere: The sum total of all life-forms on Earth and the interaction among those life-forms.

Decomposition: The breakdown of complex molecules—molecules of which dead organisms are composed—into simple nutrients that can be reutilized by living organisms.

Fossil fuel: A fuel such as coal, oil, or natural gas that is formed over millions of years from the remains of plants and animals.

Greenhouse effect: The warming of Earth's atmosphere due to water vapor, carbon dioxide, and other gases in the atmosphere that trap heat radiated from Earth's surface.

Hydrocarbons: Molecules composed solely of hydrogen and carbon atoms.

Photosynthesis: Chemical process by which plants containing chlorophyll use sunlight to manufacture their own food by converting carbon dioxide and water to carbohydrates, releasing oxygen as a by-product.

Respiration: The process in which oxygen is used to break down organic compounds into carbon dioxide and water.

same, although the carbon regularly changes its form. A particular carbon atom located in someone's eyelash may have at one time been part of some now-extinct species, like a dinosaur. Since the dinosaur died and decomposed millions of years ago, its carbon atoms have seen many forms before ending up as part of a human being. It may have been part of several plants and trees, free-floating in the air as carbon dioxide, locked away in the shell of a sea creature and then buried at the ocean bottom, or even part of a volcanic eruption. Carbon is found in great quantities in Earth's crust, its surface waters, the atmosphere, and the mass of green plants. It is also found in many different chemical combinations, including carbon dioxide (CO_2) and calcium carbonate ($CaCO_3$), as well as in a huge variety of organic compounds such as hydrocarbons (like coal, petroleum, and natural gas).

Carbon cycle processes

If a diagram were drawn showing the different processes that move carbon from one form to another, its main processes would be photosynthesis, respiration, decomposition, natural weathering of rocks, and the combustion of fossil fuels.

Photosynthesis. Carbon exists in the atmosphere as the compound carbon dioxide. It first enters the ecological food web (the connected network of producers and consumers) when photosynthetic organisms, such as plants and certain algae, absorb carbon dioxide through tiny pores in their leaves. The plants then "fix" or capture the carbon dioxide and are able to convert it into simple sugars like glucose through the biochemical process known as photosynthesis. Plants store and use this sugar to grow and to reproduce. Thus, by their very nature as makers of their own food, plants remove carbon dioxide from the atmosphere. When plants are eaten by animals, their carbon is passed on to those animals. Since animals can-

The carbon cycle. *(Reproduced by permission of The Gale Group.)*

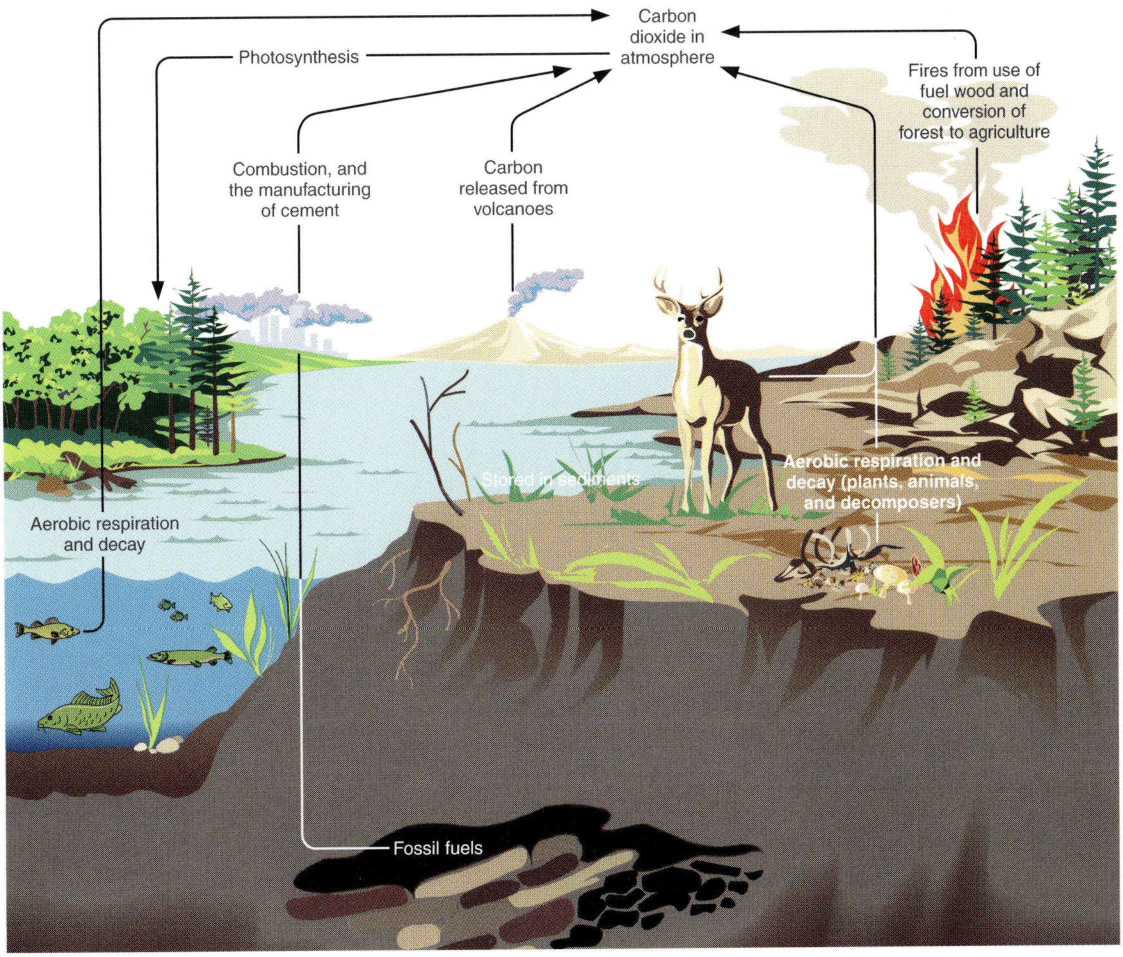

not make their own food, they must get their carbon either directly by eating plants or indirectly by eating animals that have eaten plants.

Respiration. Respiration is the next step in the cycle, and unlike photosynthesis, it occurs in plants, animals, and even decomposers. Although we usually think only of breathing oxygen when we hear the word "respiration," it has a broader meaning that involves oxygen. To a biologist, respiration is the process in which oxygen is used to break down organic compounds into carbon dioxide (CO_2) and water (H_2O). For an animal then, respiration is both taking in oxygen (and releasing carbon dioxide) and oxidizing its food (or burning it with oxygen) in order to release the energy the food contains. In both cases, carbon is returned to the atmosphere as carbon dioxide. Carbon atoms that started out as components of carbon dioxide molecules have passed through the body of living organisms and been returned to the atmosphere, ready to be recycled again.

Decomposition. Decomposition is the largest source through which carbon is returned to the atmosphere as carbon dioxide. Decomposers are microorganisms that live mostly in the soil but also in water, and which feed on the rotting remains of plants and animals. It is their job to consume both waste products and dead matter, during which they also return carbon dioxide to the atmosphere by respiration. Decomposers not only play a key role in the carbon cycle, but also break down, remove, and recycle what might be called nature's garbage.

Weathering of rocks. Not all carbon atoms are always moving somewhere in the carbon cycle. Often, many become trapped in limerock, a type of stone formed on the ocean floor by the shells of marine plankton. Sometimes after millions of years, the waters recede and the limerock is eventually exposed to the elements. When limerock is exposed to the natural process of weathering, it slowly releases the carbon atoms it contains, and they become an active part of the carbon cycle once again

Human-caused increase of carbon dioxide in the atmosphere. In recent history, humans have added to the carbon cycle by burning fossil fuels. Ever since the rapid growth of the Industrial Revolution in the nineteenth century when people first harnessed steam to power their engines, human beings have been burning carbon-containing fuels like coal and oil (called fossil fuels) for artificial power. This constant burning produces massive amounts of carbon dioxide, which are released into Earth's atmosphere. Over the last 150 years, the burning of coal, oil, and natural gas has released some 270 billion tons (245 billion metric tons) of carbon into the air in the form of carbon dioxide.

Luckily, more than half of the carbon dioxide emitted by the burning of fossil fuels is absorbed by the oceans, by plants, and by soils. Regardless, scientists feel fossil fuel consumption could be an example of a human activity that affects and possibly alters the natural processes (photosynthesis, respiration, decomposition) that nature had previously kept in balance. Many scientists believe that carbon dioxide is a "greenhouse gas." This means that it traps heat and prevents it from escaping from Earth. As a result, this trapped gas leads to a global temperature rise, a natural phenomenon known as the greenhouse effect, which can have disastrous effects on Earth's environment.

[*See also* **Greenhouse effect**]

Carbon dioxide

Carbon dioxide is a chemical compound consisting of one part carbon and two parts oxygen and represented by the chemical formula CO_2. For a number of reasons, carbon dioxide is one of the most important gases on Earth. Plants use carbon dioxide to produce carbohydrates (sugars and starches) in the process known as photosynthesis. (In photosynthesis, plants make use of light to break down chemical compounds and produce energy.) Since humans and all other animals depend on plants for their food, photosynthesis is necessary for the survival of all life on Earth.

Carbon dioxide in the atmosphere is also important because it captures heat radiated from Earth's surface. That heat keeps the planet warm enough for plant and animal (including human) life to survive. Increasing levels of carbon dioxide in the atmosphere may be responsible for long-term changes in Earth's climate. Those changes may have both beneficial and harmful effects on human and other forms of life on the planet.

History

Credit for the discovery of carbon dioxide goes to Flemish scientist Jan Baptista van Helmont (c. 1580–1644; some sources give death date as 1635). Around 1630, van Helmont identified a gas given off by burning wood and gave it the name *gas sylvestre* ("wood gas"). Today we know that gas is carbon dioxide. Van Helmont's discovery was important not only because he first recognized carbon dioxide but also because he first understood that air is a combination of gases, not a single gas.

Some of the most complete studies of carbon dioxide were conducted by Scottish chemist Joseph Black (1728–1799). In 1756, Black proved that carbon dioxide (which was then called "fixed air") occurred in the atmosphere and that it could form other compounds. He also identified carbon dioxide in the breath exhaled by humans.

The first practical use of carbon dioxide can be traced to an invention made by English chemist Joseph Priestley (1733–1804) in the mid-1700s. Priestley found that by dissolving carbon dioxide in water he could produce a fresh, sparkling beverage with a pleasant flavor. Since Priestley's discovery lacks only sugar and flavoring to make it a modern soda pop or cola drink, he can properly be called the father of the soft drink industry.

Pure carbon dioxide gas can be poured because it is heavier than air. *(Reproduced by permission of Photo Researchers, Inc.)*

Properties and uses

Carbon dioxide is a colorless, odorless, noncombustible gas with a density about twice that of air. It can be converted to a solid known as dry ice rather easily. Dry ice has the interesting property that it sublimes—that is, changes directly from a solid to a gas without first melting into a liquid. All of these properties explain the most important commercial and industrial uses of carbon dioxide.

Among those uses is the one discovered by Priestley, the manufacture of carbonated ("soft") drinks. The presence of carbon dioxide provides the slightly tart and tingly flavor that makes such beverages so refreshing. Carbon dioxide is also used widely as a coolant, a refrigerant, and an ingredient in the manufacture of frozen foods. Carbon dioxide fire extinguishers are often used to control electrical and oil fires, which cannot be put out with water. Because the gas is more dense than air and does not catch fire, it spreads like a blanket over burning material and smothers the flames. In addition, carbon dioxide is sometimes used as a gaseous blanket to prevent substances from decaying.

In rare circumstances, carbon dioxide can be a threat to life. In 1896, a huge cloud of the gas exploded from Lake Nyos, a volcanic lake in northwestern Cameroon, a nation in western Africa. The cloud spread quickly and suffocated more than 1,700 people and 8,000 animals. Today, scientists are trying to control this phenomenon by slowly pumping carbon dioxide gas from the bottom of the lake.

[*See also* **Carbon family; Combustion; Greenhouse effect; Photosynthesis; Pollution**]

Carbon family

The carbon family consists of the five elements that make up Group 14 of the periodic table: carbon, silicon, germanium, tin, and lead. The family is particularly interesting because it consists of one nonmetal (carbon), two metals (tin and lead), and two metalloids (silicon and germanium). (A metalloid is an element that has some of the properties of both metals and nonmetals.)

The atoms of all Group 14 elements have four electrons in their outermost energy level. In spite of this fact, the elements have less in common physically and chemically than do the members of most other families of elements.

Carbon

Carbon is one of the most remarkable of all chemical elements. It occurs in all living organisms. In fact, the field of organic chemistry, which began as the study of the chemistry of plants and animals, can also be called the chemistry of carbon compounds. In addition, carbon and its compounds are of critical importance to the world as sources of energy. Coal, oil, and natural gas—the so-called fossil fuels—all consist of pure carbon or carbon compounds. Finally, carbon monoxide and carbon diox-

Gem-quality diamond in Kimberlite. *(Reproduced by permission of National Aeronautics and Space Administration.)*

ide, the two oxides of carbon, are profoundly important not only in the survival of living organisms but also in a host of industrial operations. (An oxide is an inorganic compound whose only negative part is the element oxygen.)

Carbon was one of the first elements known to humans. A Greek historian of the fourth century B.C., for example, tells of a natural gas well in Turkey that provided a perpetual flame for religious ceremonies. Many reports also detail the practice of mixing lampblack, a form of carbon, with olive oil and balsam gum to make a primitive form of ink. And diamonds, another form of carbon, are described in the Bible and even older Hindu manuscripts.

Occurrence of carbon

Carbon occurs both as an element and in combined forms. As an element, it exists in at least three different allotropic forms. (Allotropes are forms of an element that differ from each other in physical and, sometimes, chemical properties.) The two best known allotropes of carbon are graphite and diamond. Graphite is a soft, shiny, dark gray or black, greasy-feeling mineral used to make the "lead" in lead pencils. Graphite is soft enough to be scratched with a fingernail.

Carbon, and a carbon-created diamond. *(Reproduced by permission of The Stock Market.)*

Buckminsterfullerene

How would you like to have a molecule named after you? That is just what happened to one famous American in 1985, when scientists discovered an entirely new form of carbon. Until that time, chemists had thought that carbon existed in only two solid forms: graphite and diamond. Then, researchers at Rice University in Texas and the University of Sussex in England found a strange-looking molecule consisting of 60 carbon atoms joined to each other in a large sphere. Under a microscope, the molecule looks like a soccer ball with 20 hexagons (six-sided figures) on its surface.

The Rice and Sussex chemists suggested naming the new molecule after American engineer and philosopher R. Buckminster Fuller (1895–1983). Fuller had created a number of exciting new architectural forms, one of which was the geodesic dome. A geodesic dome, like the new molecule, is a sphere made of many plane (flat) figures like the hexagon. Because of this similarity, the new molecule was given the name buckminsterfullerene or, more briefly, fullerene. Less formally, the molecules are also known as bucky-balls.

The discovery of fullerenes has created a whole new field of chemistry that involves the study of "hollow" molecules in the shape of spheres or cylindrical rods. In the early 1990s, astronomers announced the discovery of fullerene molecules in outer space.

The second common allotrope of carbon is diamond. In striking contrast with graphite, diamond is the world's hardest natural material. Its ability to bend and spread light produces the spectacular rainbow "fire" that is often associated with diamond jewelry. Skillful gem cutters are able to cut and polish diamonds in a way that maximizes the effect of this natural property.

In 1985, a third allotropic form of carbon was discovered. It is a 60-atom structure called buckminsterfullerene that looks like a soccer ball when viewed under a microscope.

Carbon also occurs in a number of common compounds. Carbon dioxide, for example, is the fifth most abundant gas in the atmosphere. It makes up about 0.3 percent of the total volume of all atmospheric gases. Calcium carbonate is one of the most abundant of all rocks in Earth's crust. It occurs in a wide variety of forms, such as limestone, marble, travertine, chalk, and dolomite. Stalactites and stalagmites in caves are made of calcium carbonate, as are many animal products, such as coral, sea shells, egg shells, and pearls.

Carbon exists abundantly in Earth's crust in the form of the fossil fuels. The fossil fuels are coal, oil, and natural gas. They have been given this name because they were apparently formed—in the absence of oxygen—by the slow decay of plant and animal forms that lived millions of years ago.

Why carbon is special

More than ten million compounds of carbon are now known. That number is far greater than the total of all noncarbon compounds that have been discovered. The special property that makes carbon so different from all other elements is the ability of its atoms to combine with each other in long chains. It is possible to find compounds in which two atoms of an element are joined to each other, but chains of more than two are rare. A chain of ten or more atoms (other than carbon) is virtually unheard of. Yet long chains of carbon atoms are the rule rather than the exception. For example, the protein molecules in your body consist of hundreds or thousands of carbon atoms connected to each other in a long chain.

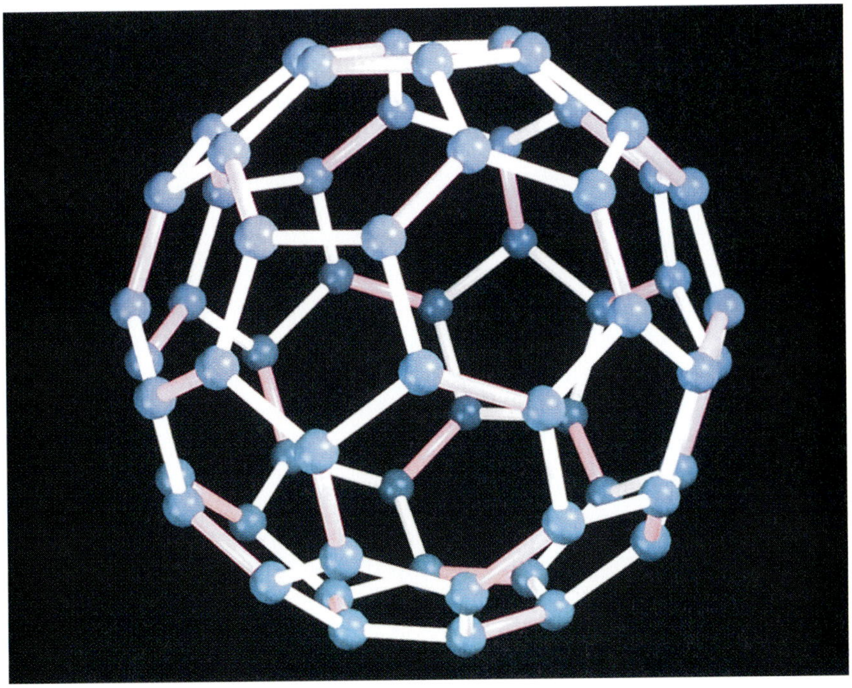

A supercomputer simulation of the atomic structure of a molecule of Buckminster-fullerene. Carbon molecules appear as small spheres. *(Reproduced by permission of Photo Researchers, Inc.)*

(Proteins are large molecules that are essential to the structure and functioning of all living cells.)

Furthermore, carbon atoms can form structures more complicated than chains. Some compounds have carbon chains with other chains branching off from them, carbon chains joined tail-to-end in rings or rings inside of rings, carbon chains in the shape of cages, boxes, and spheres, and carbon chains in other strange and fascinating shapes. The interesting point is that these strange molecular structures are not just laboratory curiosities. In many cases, they are found in some of the most important compounds in living organisms.

Silicon

Silicon is the second most abundant element in Earth's crust after oxygen. It makes up about 27.2 percent of the crust by weight. In the universe as a whole, silicon is the seventh most abundant element, after hydrogen, helium, carbon, nitrogen, oxygen, and neon.

Silicon never occurs free in nature; except for rare occasions, it is found in combination with oxygen. Probably the best-known and most important mineral of silicon is silicon dioxide. Silicon dioxide occurs in two forms: a crystalline form known as quartz and a noncrystalline form known as flint. Quartz itself appears in a variety of forms, perhaps the best known of which is sand.

Other than silicon dioxide, the most common forms of silicon in nature are silicates, compounds that contain silicon, oxygen, and other elements such as magnesium, aluminum, calcium, and phosphorus. Some familiar silicates include asbestos, beryl, various types of clay, feldspar, mica, talc, and the vast majority of gemstones, including emeralds, sapphires, and rubies. The commercial product known as Portland cement also consists largely of silicates.

Uses. Various minerals of silicon have long been used in human societies for two purposes: in construction and in the manufacture of glasses and ceramics. Many natural building materials—brick, clay, and sandstone, for instance—are made of or contain silicate materials. And a very wide variety of glasses and ceramics are made by combining silicon dioxide or silicates with other blending materials.

Silicon itself also has many important uses. It was one of the first materials to be used in the manufacture of semiconductors and is still in demand for that application. (Semiconductors are substances that conduct an electric current but do so very poorly.) It is also used as an alloying agent with metals, particularly to increase the strength of a given alloy.

Another application of silicon is in the manufacture of certain plasticlike materials known as silicones. Silicones consist of silicon combined with a variety of organic materials. Low-molecular-weight silicones are oils that make good lubricants; they stand up better to high and low temperatures than do petroleum oils. High-molecule-weight silicones are known as silicone rubber. Silicone rubber is used in a variety of ways—in everything from super-bouncing balls to bathtub sealers to spaceship parts. The first human footprint on the Moon was made with a silicone-rubber-soled boot. In between silicone oils and rubber are hundreds of kinds of silicones used in electrical insulators, rust preventives, soaps, fabric softeners, hair sprays, hand creams, furniture and auto polishes, paints, adhesives, and chewing gum.

Germanium

Germanium has earned a place in the history of chemistry as the element predicted to exist by Russian chemist Dmitry Mendeleev (1834–1907) in 1871. Mendeleev's ability to predict the properties of the as-yet-unknown element simply because of its place in the periodic table was a remarkable accomplishment. Not only did it lead to the search for the element but it also convinced chemists that Mendeleev's scheme for organizing the elements was a valid one.

Germanium metal is a dark gray solid with a metallic luster. Its melting point is 937.4°C (1,719°F), and its boiling point is 2,830°C (5,130°F). Germanium sometimes occurs as a pure metal in combination with more abundant elements and as germanium oxide. The element is not very abundant in Earth's surface, occurring to the extent of no more than 1 to 2 ppm (parts per million).

By far the most important application of germanium is in the manufacture of semiconductors. The addition of small amounts of impurities such as arsenic, phosphorus, and gallium greatly increases silicon's conductivity and its value in electronic devices such as transistors.

Tin

Tin is one of the first metals to have been used by humans. The earliest written records date to about 3500 B.C., when tools and weapons made of bronze (an alloy of tin and copper) were in general use. In fact, the success of bronze for these applications gave the period a name by which it is now well known, the Bronze Age.

Tin is not a particularly abundant element in Earth's crust, ranking forty-eighth among the elements. It follows bromine and uranium and precedes beryllium and arsenic in order of abundance. It occurs primarily in the form of the mineral cassiterite, an oxide of tin.

Tin occurs in two allotropic forms at and near room temperature. One form is a silvery-white metallic material that is both ductile (capable of being drawn out into a fine wire) and malleable (capable of being rolled or hammered into thin sheets). It is called white tin and has a melting point of 232°C (450°F) and a boiling point of 2,260°C (4,100°F). The other form is a brittle powdery solid known as gray tin. An interesting illustration of these two forms of tin occurred in 1850, when temperatures in Russia dropped below freezing for an extended period. Objects made of tin—including buttons, organ pipes, and drinking cups—changed from stable white tin to fragile gray tin. These objects crumbled and fell apart as temperatures continued to hover around the freezing point.

Uses. Tin is too soft and fragile to be used by itself. Its applications almost always involve a blending of the metal with one or more other metals. For example, its most important application is in tin-plating other metals. Tin is applied to the surface of iron or steel to provide a nontoxic, corrosion-resistant product. Alloys in which tin is used include solder, bronze, Babbitt metal, pewter, and type metals.

Lead

Lead is another metal that has been known to humans for thousands of years. It was used for making pottery glazes in Egypt as early as the seventh millennium B.C., as roofing and flooring in Babylonia, and for water pipes and other types of plumbing in ancient Rome.

This metal is by far the most abundant of the heavy metals with an occurrence of about 13 parts per million in Earth's crust. It occurs most commonly as the black mineral galena (lead sulfide). It also occurs in angelsite (lead sulfate), cerussite (lead carbonate), pyromorphite (lead phosphochloride), and mimetesite (lead arsenochloride).

Lead is a dull, gray, soft metal with a density of 11.342 grams per cubic centimeter. Its density is about twice that of tin and five times that of silicon. Its melting point is 327°C (621°F), and its boiling point is 1,751°C (3,184°F).

Uses. The uses of lead have changed dramatically over the past half century because of discoveries about its toxicity. At one time, lead was used in many applications that involved its entering the human body. For example, household plumbing devices and many kinds of pots and pans contained lead. When water passed through these devices or the pots and pans were used for cooking, small amounts of lead were dissolved and were consumed by humans.

Today we know that lead has serious effects on the human body. It can cause damage to the liver, kidneys, and brain. It is especially harmful to young children whose mental development can be severely impaired by the consumption of lead-containing materials.

These health effects have dramatically changed the pattern of lead use in the United States and many other parts of the world. In the first half of the twentieth century, lead was widely used in house paints and in leaded gasoline. Today, both applications of lead have been banned in an attempt to reduce the exposure of humans to the metal.

Instead, more than half of all the lead produced today goes to the manufacture of lead storage batteries used in cars, trucks, and other vehicles. The lead is used to produce the two electrodes used in such batteries. The disposal of lead storage batteries (once they are no longer usable) continues to be an ongoing environmental problem in many parts of the world.

Lead can also be found in a variety of alloys, such as solder and type metal; limited-use paints and pigments; in certain types of glasses and ceramics; in the production of special-use pipes, tubes, sheeting, and lead foil; and in protective coatings for cables.

[*See also* **Alloy; Element, chemical; Organic chemistry; Periodic table; Transistor**]

Carbon monoxide

Carbon monoxide is a compound of carbon and oxygen in which the ratio of the two elements is one atom of carbon to one atom of oxygen. Its formula is CO. Carbon monoxide is a colorless, odorless, tasteless, poisonous gas. Most people have heard about carbon monoxide because of its toxic effects. People who live or work in crowded urban areas may become ill with headaches and nausea because of exposure to carbon monoxide in polluted air. In higher concentrations, the gas can even cause death.

History

The early history of gases such as carbon monoxide is sometimes difficult to trace. Until the early 1600s, scientists did not realize that the material we call air is actually a mixture of gases. As early as the late

Words to Know

Combustion: Oxidation that occurs so rapidly that noticeable heat and light are produced; burning.

Hemoglobin: An complex iron-containing molecule that transports oxygen through the circulatory system.

Incomplete combustion: Combustion that occurs in such a way that fuel is not completely oxidized ("burned up"). The incomplete combustion of carbon-containing fuels (such as coal and oil) always results in the formation of some carbon monoxide.

Reducing agent: A substance that removes oxygen from some other material.

Toxic: Poisonous.

thirteenth century, Spanish alchemist Arnold of Villanova (c. 1235–1311) described a poisonous gas formed by the burning of wood; this gas was almost certainly carbon monoxide.

Flemish scientist Jan Baptista van Helmont (c. 1580–1644; some sources give death date as 1635) nearly died as a result of inhaling *gas carbonum,* apparently a mixture of carbon monoxide and carbon dioxide. Credit for the discovery of carbon monoxide, however, is usually given to English chemist and theologian Joseph Priestley (1733–1804). During the period between 1772 and 1799, Priestley gradually recognized the difference between carbon dioxide and carbon monoxide and correctly stated the properties of the latter gas.

Sources

Like carbon dioxide, carbon monoxide is formed naturally during the combustion (burning) of wood, coal, and other naturally occurring substances. Huge quantities of carbon monoxide are produced, for example, during a forest fire or a volcanic eruption.

The relative amounts of carbon monoxide or carbon dioxide that form during combustion depend on two factors: the amount of oxygen present and the combustion temperature. When a large supply of oxygen is present and when the combustion temperature is high, carbon dioxide is more likely to be formed. With limited supplies of oxygen and at lower temperatures, carbon monoxide is produced.

Carbon monoxide is not extracted from the air very easily but is produced commercially by the controlled oxidation of carbon. For example, producer gas is a product made by blowing air across very hot coke (nearly pure carbon). Producer gas consists of three gases: carbon monoxide, carbon dioxide, and nitrogen in the ratio of 6 to 1 to 18. Water gas is made by a similar process—passing steam over hot coke. The products in this case are hydrogen, carbon monoxide, carbon dioxide, and other gases in the ration of 10 to 8 to 1 to 1.

Physiological effects

The poisonous character of carbon monoxide has been well known for many centuries. At low concentrations, carbon monoxide may cause nausea, vomiting, restlessness, and euphoria (a feeling of well-being). As exposure increases, a person may lose consciousness and go into convulsions. Death is a common final result. The U.S. Occupational Safety and Health Administration has established a limit of 35 parts per million of carbon monoxide in workplaces where a person may be continually exposed to the gas.

Scientists now know how carbon monoxide poisoning occurs. Normally, oxygen is transported from the lungs to cells by means of red blood cells. This process occurs when oxygen atoms bond to an iron atom in the middle of a complex molecule known as oxyhemoglobin. Oxyhemoglobin is a fairly unstable molecule that breaks down to release free oxygen and hemoglobin for use by the body's cells. The oxygen is then available to carry out reactions in cells from which the body gets energy.

If carbon monoxide is present in the lungs, this sequence of reactions is disrupted. Carbon monoxide bonds with iron in hemoglobin to form carbonmonoxyhemoglobin, a complex somewhat similar to oxyhemoglobin. Carbonmonoxyhemoglobin, however, is a more stable compound than oxyhemoglobin. When it reaches cells, it has little tendency to break apart; instead, it continues to circulate in the bloodstream in its bound form. As a result, cells are unable to obtain the oxygen they need for energy production, and the symptoms of carbon monoxide poisoning begin to appear.

Carbon monoxide poisoning—at least at moderate levels—is so common in everyday life that carbon monoxide detectors, similar to smoke alarms, are found in many businesses and homes. Poorly ventilated charcoal fires, improperly installed gas appliances, and exhaust from automobiles and trucks are the most common sources of the gas. In fact, levels of carbon monoxide in the air can become dangerously high in busy urban areas that have large numbers of cars and trucks. Cigarette smokers may also be exposed to harmful levels of the gas. Studies have shown that the one-to-two pack-a-day smoker may have up to 7 percent of the hemoglobin in her or his body tied up in the form of carbonmonoxyhemoglobin.

Uses

Carbon monoxide is used in industry primarily as a source of energy and as a reducing agent. Both producer and water gas are burned as fuels for a variety of industrial operations. As a reducing agent, carbon monoxide is used to convert the naturally occurring oxide of a metal to the pure metal. When carbon monoxide is passed over hot iron oxides, for example, the oxides are converted to metallic iron.

[*See also* **Carbon dioxide; Carbon family**]

Carcinogen

A carcinogen is a substance that causes a normal cell to change into a cancerous cell, resulting in uncontrolled cell growth. Cancer cells can multiply unchecked, forming a mass of tissue called a tumor. Sometimes cancerous cells "break off" from a tumor, traveling through the body and multiplying in other organs and tissues.

How carcinogens cause cancer

Carcinogens cause cancer by producing changes (or mutations) in the genetic material, or DNA, of a cell. These mutations result in uncontrolled cell division. A cancer-causing substance can alter the DNA of a cell directly or it can react with other chemicals in the body to form substances that cause gene mutations. (Genes are sections of DNA that serve as units of hereditary information.) A transformed cell may continue to function normally and not begin to multiply and develop into a tumor until many months or years later. It is not unusual for cancer to appear 20 to 25 years after initial exposure to a carcinogen.

Types of carcinogens

Carcinogens include both naturally occurring and artificially produced chemicals, ultraviolet light, and radioactive substances such as radon (a radioactive gas that is present in rock).

Words to Know

Aflatoxin: A carcinogenic poison produced by a mold that grows on peanuts and grains.

Cancer: A disease of uncontrolled cell growth.

DNA (deoxyribonucleic acid): The genetic material in the nucleus of cells that contains information for an organism's development.

Mutation: A change in the genetic material of a cell.

Radiation: Energy that is sent out as waves or particles.

Tumor: A mass of tissue formed by abnormal cell growth.

About 23 chemicals have been identified as carcinogens in humans, with many more shown to cause cancer in laboratory animals. Tobacco smoke contains several carcinogenic substances and is the major cause of lung cancer. Some drugs used in the treatment of cancer are themselves

A hazardous waste warning limits travel through Times Beach, Missouri. *(Reproduced by permission of Greenpeace Photos.)*

Ames Test

The Ames test is a quick method of determining if a substance is capable of producing mutations. A culture of a strain of Salmonella bacteria that lacks an enzyme needed for growth is exposed to possible carcinogens. If the substance added to the culture is carcinogenic, it will cause mutations in the bacteria that allow the bacteria to grow. The Ames test has positively identified many carcinogens. It is used by cosmetic companies, drug manufacturers, and other industries that must prove that their products will not cause cancer in humans.

cancer-causing. In addition, various chemicals used in industrial processes, such as vinyl chloride and certain dyes, are known human carcinogens. Nitrites—chemicals added to processed meats such as bacon, sausage, and bologna to prevent spoilage—react with substances in the digestive tract to form carcinogenic chemicals called nitrosamines. Even synthetic dyes added to food for coloring are potential mutagens (substances that cause mutations in the genetic material of cells).

Sunlight is a well-known carcinogen that can cause changes in skin cells that may lead to skin cancer. Radiation emitted from an atomic bomb or released in nuclear power accidents can result in cancer in people exposed to it. Repeated exposure to radiation from medical X rays or other sources also may increase a person's risk of developing genetic mutations.

Some foods, such as celery, black pepper, white mushrooms, and mustard contain naturally occurring carcinogens. Aflatoxin is a cancer-causing chemical produced by molds on peanuts. However, these foods must be consumed in large quantities over a long period to initiate cancer.

[*See also* **Cancer; Cigarette smoke; Mutation; Nucleic acid; Radiation exposure; Virus**]

Carpal tunnel syndrome

Carpal tunnel syndrome is a condition in which the squeezing or compressing of a nerve that passes through the wrist results in numbness, tingling, weakness, or pain in one or both hands. The hands may become so weakened that opening jars or grasping objects becomes difficult and painful.

The carpal tunnel is a space formed by the carpal (wrist) bones and the carpal ligament (a connective tissue that attaches bone to bone). Through this space pass the median nerve and tendons of the fingers and thumb. (The median nerve runs from the neck through the middle of the arm to the fingers. Tendons are connective tissue that attach muscle to bone.) When the tendons within the carpal tunnel become inflamed, they swell and press on the median nerve.

Continuous typing on a keyboard can cause carpal tunnel syndrome. *(Reproduced by permission of Field Mark Publications.)*

Causes of median nerve compression

A number of conditions can cause swelling of the carpal tunnel, leading to pressure on the median nerve. Such conditions include pregnancy, arthritis (inflammation of the joints), hypothyroidism (reduced function of the thyroid, a gland located in the neck that plays an important role in metabolism, or the conversion of food to energy), diabetes (an inability to metabolize—or break down—sugar properly), menopause (the point in a woman's life when menstruation ceases and childbearing is no longer possible), and pituitary abnormalities. (The pituitary or master gland secretes substances that directly or indirectly influence most basic bodily functions). Also, performing a job that requires repeated bending or twisting of the wrists increases the likelihood of developing the disorder. Continuous flexing of the wrist, as when typing on a keyboard or playing a piano, can cause compression of the median nerve. Carpal tunnel syndrome is much more common among women than men.

Treatment

Carpal tunnel syndrome is treated initially by applying a brace, or splint, to prevent the wrist from bending and to relieve pressure on the median nerve. If a person's job is causing the disorder, performing other work may be necessary. Treatment of a related medical condition may relieve the symptoms of carpal tunnel syndrome. Severe cases may require surgery to decrease compression of the median nerve.

Cartography

Cartography is the art of making maps. A map is a two-dimensional (flat) drawing or chart showing the political boundaries and physical features of a geographical region. For example, a map may show the location of cities, mountain ranges, and rivers, or may show the types of rock in a given region. Cartography is considered a subdiscipline of geography, which is the study of Earth's surface and its various climates, continents, countries, and resources.

The history of cartography

The oldest known map is of an area in northern Mesopotamia, an ancient region in southwest Asia. The baked clay tablet, found near present-day Nuzi, Iraq, dates from approximately 3800 B.C. Fragments of clay maps nearly 4,000 years old have been found in other parts of

Mesopotamia, some showing city plans and others showing parcels of land. Over 3,000 years ago, the ancient Egyptians surveyed the lands in the Nile Valley. They drew detailed maps on papyrus for use in taxation.

The ancient Greeks developed many of the basic principles of modern cartography, including latitude and longitude, and map projections. The maps of Ptolemy, a Greek astronomer and mathematician who lived in the second century A.D., are considered the high point of Greek cartography.

The era of European exploration that arose in the 1500s supplied cartographers with a wealth of new information, which allowed them to produce maps and navigation charts of ever-increasing accuracy and detail. Europeans became fascinated with the idea of mapping the world. The French initiated the first national survey during the 1700s, and soon other European countries followed suit. Today, most countries have an official organization devoted to cartographic research and production.

Mapmaking

No single map can accurately show every feature on Earth's surface. There is simply too much spatial information at any particular point

A relief map of the United States.

for all of the information to be presented in a comprehensible, usable format. Maps show the location of selected phenomena by using symbols that are identified in a legend.

Maps are smaller than the area they depict: they present various pieces of geographical information at a reduced scale. Every map has a statement of its scale, which is an expression of the ratio between map distance and actual distance. This statement can take many forms, and many maps express scale in more than one way. The graphic scale is a line or bar showing how many actual miles or kilometers are represented by a particular number of inches or centimeters on the map. Another scale indicates distance as a ratio between two points on the map and their actual geographical distance. For example, a map with a scale of 1:100,000 tells the map reader that every 1 unit of distance on the map equals 100,000 of the same units of distance on the ground. In this example, 1 inch or centimeter on the map would equal 100,000 inches or centimeters in actual ground distance.

Cartographers traditionally obtained their information from navigators and surveyors. For many centuries maps were produced entirely by hand. They were drawn or painted on paper, hide, parchment, clay tablets, and slabs of wood. Each map was an original work. Once the printing press was developed in Europe in the 1400s, many reproductions were made from an original work. Maps became more common and more accessible.

Various techniques were integrated into the printing process during the last 200 years, increasing the variety of scales at which maps were produced. The introduction of the lithographic printing method in the late 1800s brought about the production of multicolored maps. Today, cartographers incorporate information from aerial photography and satellite imagery in the maps they create. They also use computer-assisted design programs to produce map images.

[*See also* **Geologic map**]

Where to Learn More

Books

Earth Sciences

Cox, Reg, and Neil Morris. *The Natural World.* Philadelphia, PA: Chelsea House, 2000.

Dasch, E. Julius, editor. *Earth Sciences for Students.* Four volumes. New York: Macmillan Reference, 1999.

Denecke, Edward J., Jr. *Let's Review: Earth Science.* Second edition. Hauppauge, NY: Barron's, 2001.

Engelbert, Phillis. *Dangerous Planet: The Science of Natural Disasters.* Three volumes. Farmington Hills, MI: UXL, 2001.

Gardner, Robert. *Human Evolution.* New York: Franklin Watts, 1999.

Hall, Stephen. *Exploring the Oceans.* Milwaukee, WI: Gareth Stevens, 2000.

Knapp, Brian. *Earth Science: Discovering the Secrets of the Earth.* Eight volumes. Danbury, CT: Grolier Educational, 2000.

Llewellyn, Claire. *Our Planet Earth.* New York: Scholastic Reference, 1997.

Moloney, Norah. *The Young Oxford Book of Archaeology.* New York: Oxford University Press, 1997.

Nardo, Don. *Origin of Species: Darwin's Theory of Evolution.* San Diego, CA: Lucent Books, 2001.

Silverstein, Alvin, Virginia Silverstein, and Laura Silverstein Nunn.*Weather and Climate.* Brookfield, CN: Twenty-First Century Books, 1998.

Williams, Bob, Bob Ashley, Larry Underwood, and Jack Herschbach. *Geography.* Parsippany, NJ: Dale Seymour Publications, 1997.

Life Sciences

Barrett, Paul M. *National Geographic Dinosaurs.* Washington, D.C.: National Geographic Society, 2001.

Fullick, Ann. *The Living World.* Des Plaines, IL: Heinemann Library, 1999.

Gamlin, Linda. *Eyewitness: Evolution.* New York: Dorling Kindersley, 2000.

Greenaway, Theresa. *The Plant Kingdom: A Guide to Plant Classification and Biodiversity.* Austin, TX: Raintree Steck-Vaughn, 2000.

Kidd, J. S., and Renee A Kidd. *Life Lines: The Story of the New Genetics.* New York: Facts on File, 1999.

Kinney, Karin, editor. *Our Environment.* Alexandria, VA: Time-Life Books, 2000.

Nagel, Rob. *Body by Design: From the Digestive System to the Skeleton.* Two volumes. Farmington Hills, MI: UXL., 2000.

Parker, Steve. *The Beginner's Guide to Animal Autopsy: A "Hands-in" Approach to Zoology, the World of Creatures and What's Inside Them.* Brookfield, CN: Copper Beech Books, 1997.

Pringle, Laurence. *Global Warming: The Threat of Earth's Changing Climate.* New York: SeaStar Books, 2001.

Riley, Peter. *Plant Life.* New York: Franklin Watts, 1999.

Stanley, Debbie. *Genetic Engineering: The Cloning Debate.* New York: Rosen Publishing Group, 2000.

Whyman, Kate. *The Animal Kingdom: A Guide to Vertebrate Classification and Biodiversity.* Austin, TX: Raintree Steck-Vaughn, 1999.

Physical Sciences

Allen, Jerry, and Georgiana Allen. *The Horse and the Iron Ball: A Journey Through Time, Space, and Technology.* Minneapolis, MN: Lerner Publications, 2000.

Berger, Samantha, *Light.* New York: Scholastic, 1999.

Bonnet, Bob L., and Dan Keen. *Physics.* New York: Sterling Publishing, 1999.

Clark, Stuart. *Discovering the Universe.* Milwaukee, WI: Gareth Stevens, 2000.

Fleisher, Paul, and Tim Seeley. *Matter and Energy: Basic Principles of Matter and Thermodynamics.* Minneapolis, MN: Lerner Publishing, 2001.

Gribbin, John. *Eyewitness: Time and Space.* New York: Dorling Kindersley, 2000.

Holland, Simon. *Space.* New York: Dorling Kindersley, 2001.

Kidd, J. S., and Renee A. Kidd. *Quarks and Sparks: The Story of Nuclear Power.* New York: Facts on File, 1999.

Levine, Shar, and Leslie Johnstone. *The Science of Sound and Music.* New York: Sterling Publishing, 2000

Naeye, Robert. *Signals from Space: The Chandra X-ray Observatory.* Austin, TX: Raintree Steck-Vaughn, 2001.

Newmark, Ann. *Chemistry.* New York: Dorling Kindersley, 1999.

Oxlade, Chris. *Acids and Bases.* Chicago, IL: Heinemann Library, 2001.

Vogt, Gregory L. *Deep Space Astronomy.* Brookfield, CT: Twenty-First Century Books, 1999.

Technology and Engineering Sciences

Baker, Christopher W. *Scientific Visualization: The New Eyes of Science.* Brookfield, CT: Millbrook Press, 2000.

Cobb, Allan B. *Scientifically Engineered Foods: The Debate over What's on Your Plate.* New York: Rosen Publishing Group, 2000.

Cole, Michael D. *Space Launch Disaster: When Liftoff Goes Wrong.* Springfield, NJ: Enslow, 2000.

Deedrick, Tami. *The Internet.* Austin, TX: Raintree Steck-Vaughn, 2001.

DuTemple, Leslie A. *Oil Spills.* San Diego, CA: Lucent Books, 1999.

Gaines, Ann Graham. *Satellite Communication.* Mankata, MN: Smart Apple Media, 2000.

Gardner, Robert, and Dennis Shortelle. *From Talking Drums to the Internet: An Encyclopedia of Communications Technology.* Santa Barbara, CA: ABC-Clio, 1997.

Graham, Ian S. *Radio and Television.* Austin, TX: Raintree Steck-Vaughn, 2000.

Parker, Steve. *Lasers: Now and into the Future.* Englewood Cliffs, NJ: Silver Burdett Press, 1998.

Sachs, Jessica Snyder. *The Encyclopedia of Inventions.* New York: Franklin Watts, 2001.

Wilkinson, Philip. *Building.* New York: Dorling Kindersley, 2000.

Wilson, Anthony. *Communications: How the Future Began.* New York: Larousse Kingfisher Chambers, 1999.

Periodicals

Archaeology. Published by Archaeological Institute of America, 656 Beacon Street, 4th Floor, Boston, Massachusetts 02215. Also online at www.archaeology.org.

Astronomy. Published by Kalmbach Publishing Company, 21027 Crossroads Circle, Brookfield, WI 53186. Also online at www.astronomy.com.

Discover. Published by Walt Disney Magazine, Publishing Group, 500 S. Buena Vista, Burbank, CA 91521. Also online at www.discover.com.

National Geographic. Published by National Geographic Society, 17th & M Streets, NW, Washington, DC 20036. Also online at www.nationalgeographic.com.

New Scientist. Published by New Scientist, 151 Wardour St., London, England W1F 8WE. Also online at www.newscientist.com (includes links to more than 1,600 science sites).

Popular Science. Published by Times Mirror Magazines, Inc., 2 Park Ave., New York, NY 10024. Also online at www.popsci.com.

Science. Published by American Association for the Advancement of Science, 1333 H Street, NW, Washington, DC 20005. Also online at www.sciencemag.org.

Science News. Published by Science Service, Inc., 1719 N Street, NW, Washington, DC 20036. Also online at www.sciencenews.org.

Scientific American. Published by Scientific American, Inc., 415 Madison Ave, New York, NY 10017. Also online at www.sciam.com.

Smithsonian. Published by Smithsonian Institution, Arts & Industries Bldg., 900 Jefferson Dr., Washington, DC 20560. Also online at www.smithsonianmag.com.

Weatherwise. Published by Heldref Publications, 1319 Eighteenth St., NW, Washington, DC 20036. Also online at www.weatherwise.org.

Web Sites

Cyber Anatomy (provides detailed information on eleven body systems and the special senses) *http://library.thinkquest.org/11965/*

The DNA Learning Center (provides in-depth information about genes for students and educators) *http://vector.cshl.org/*

Educational Hotlists at the Franklin Institute (provides extensive links and other resources on science subjects ranging from animals to wind energy) *http://sln.fi.edu/tfi/hotlists/hotlists.html*

ENC Web Links: Science (provides an extensive list of links to sites covering subject areas under earth and space science, physical science, life science, process skills, and the history of science) *http://www.enc.org/weblinks/science/*

ENC Web Links: Math topics (provides an extensive list of links to sites covering subject areas under topics such as advanced mathematics, algebra, geometry, data analysis and probability, applied mathematics, numbers and operations, measurement, and problem solving) *http://www.enc.org/weblinks/math/*

Encyclopaedia Britannica Discovering Dinosaurs Activity Guide *http://dinosaurs.eb.com/dinosaurs/study/*

The Exploratorium: The Museum of Science, Art, and Human Perception *http://www.exploratorium.edu/*

Where to Learn More

ExploreMath.com (provides highly interactive math activities for students and educators) *http://www.exploremath.com/*

ExploreScience.com (provides highly interactive science activities for students and educators) *http://www.explorescience.com/*

Imagine the Universe! (provides information about the universe for students aged 14 and up) *http://imagine.gsfc.nasa.gov/*

Mad Sci Network (highly searchable site provides extensive science information in addition to a search engine and a library to find science resources on the Internet; also allows students to submit questions to scientists) *http://www.madsci.org/*

The Math Forum (provides math-related information and resources for elementary through graduate-level students) *http://forum.swarthmore.edu/*

NASA Human Spaceflight: International Space Station (NASA homepage for the space station) *http://www.spaceflight.nasa.gov/station/*

NASA's Origins Program (provides up-to-the-minute information on the scientific quest to understand life and its place in the universe) *http://origins.jpl.nasa.gov/*

National Human Genome Research Institute (provides extensive information about the Human Genome Project) *http://www.nhgri.nih.gov:80/index.html*

New Scientist Online Magazine *http://www.newscientist.com/*

The Nine Planets (provides a multimedia tour of the history, mythology, and current scientific knowledge of each of the planets and moons in our solar system) *http://seds.lpl.arizona.edu/nineplanets/nineplanets/nineplanets.html*

The Particle Adventure (provides an interactive tour of quarks, neutrinos, antimatter, extra dimensions, dark matter, accelerators, and particle detectors) *http://particleadventure.org/*

PhysLink: Physics and astronomy online education and reference *http://physlink.com/*

Savage Earth Online (online version of the PBS series exploring earthquakes, volcanoes, tsunamis, and other seismic activity) *http://www.pbs.org/wnet/savageearth/*

Science at NASA (provides breaking information on astronomy, space science, earth science, and biological and physical sciences) *http://science.msfc.nasa.gov/*

Science Learning Network (provides Internet-guided science applications as well as many middle school science links) *http://www.sln.org/*

SciTech Daily Review (provides breaking science news and links to dozens of science and technology publications; also provides links to numerous "interesting" science sites) *http://www.scitechdaily.com/*

Space.com (space news, games, entertainment, and science fiction) *http://www.space.com/index.html*

SpaceDaily.com (provides latest news about space and space travel) *http://www.spacedaily.com/*

SpaceWeather.com (science news and information about the Sun-Earth environment) *http://www.spaceweather.com/*

The Why Files (exploration of the science behind the news; funded by the National Science Foundation) *http://whyfiles.org/*

Index

Italic type indicates volume numbers; **boldface** type indicates entries and their page numbers; (ill.) indicates illustrations.

A

Abacus *1:* **1-2** 1 (ill.)
Abelson, Philip *1:* 24
Abortion *3:* 565
Abrasives *1:* **2-4,** 3 (ill.)
Absolute dating *4:* 616
Absolute zero *3:* 595-596
Abyssal plains *7:* 1411
Acceleration *1:* **4-6**
Acetylsalicylic acid *1:* **6-9,** 8 (ill.)
Acheson, Edward G. *1:* 2
Acid rain *1:* **9-14,** 10 (ill.), 12 (ill.), *6:* 1163, *8:* 1553
Acidifying agents *1:* 66
Acids and bases *1:* **14-16,** *8:* 1495
Acoustics *1:* **17-23,** 17 (ill.), 20 (ill.)
Acquired immunodeficiency syndrome. *See* **AIDS (acquired immunodeficiency syndrome)**
Acrophobia *8:* 1497
Actinides *1:* **23-26,** 24 (ill.)
Acupressure *1:* 121
Acupuncture *1:* 121
Adams, John Couch *7:* 1330
Adaptation *1:* **26-32,** 29 (ill.), 30 (ill.)
Addiction *1:* **32-37,** 35 (ill.), *3:* 478
Addison's disease *5:* 801
Adena burial mounds *7:* 1300
Adenosine triphosphate *7:* 1258
ADHD *2:* 237-238
Adhesives *1:* **37-39,** 38 (ill.)
Adiabatic demagnetization *3:* 597
ADP *7:* 1258
Adrenal glands *5:* 796 (ill.)
Adrenaline *5:* 800
Aerobic respiration *9:* 1673
Aerodynamics *1:* **39-43,** 40 (ill.)
Aerosols *1:* **43-49,** 43 (ill.)
Africa *1:* **49-54,** 50 (ill.), 53 (ill.)
Afterburners *6:* 1146
Agent Orange *1:* **54-59,** 57 (ill.)
Aging and death *1:* **59-62**
Agoraphobia *8:* 1497
Agriculture *1:* **62-65,** 63, 64 (ill.), *3:*582-590, *5:* 902-903, *9:* 1743-744, *7:* 1433 (ill.)
Agrochemicals *1:* **65-69,** 67 (ill.), 68 (ill.)
Agroecosystems *2:* 302
AI. *See* **Artificial intelligence**
AIDS (acquired immunodeficiency syndrome) *1:* **70-74,** 72 (ill.), *8:* 1583, *9:* 1737
Air flow *1:* 40 (ill.)
Air masses and fronts *1:* **80-82,** 80 (ill.)
Air pollution *8:* 1552, 1558
Aircraft *1:* **74-79,** 75 (ill.), 78 (ill.)
Airfoil *1:* 41
Airplanes. *See* **Aircraft**
Airships *1:* 75

Index

Al-jabr wa'l Muqabalah 1: 97
Al-Khwarizmi *1:* 97
Alchemy *1:* **82-85**
Alcohol (liquor) *1:* 32, 85-87
Alcoholism *1:* **85-88**
Alcohols *1:* **88-91,** 89 (ill.)
Aldrin, Edwin *9:* 1779
Ale *2:* 354
Algae *1:* **91-97,** 93 (ill.), 94 (ill.)
Algal blooms *1:* 96
Algebra *1:* **97-99,** *2:* 333-334
Algorithms *1:* 190
Alkali metals *1:* **99-102,** 101 (ill.)
Alkaline earth metals *1:* **102-106,** 104 (ill.)
Alleles *7:* 1248
Allergic rhinitis *1:* 106
Allergy *1:* **106-110,** 108 (ill.)
Alloy *1:* **110-111**
Alpha particles *2:* 233, *8:* 1620, 1632
Alps *5:* 827, *7:* 1301
Alternating current (AC) *4:* 741
Alternation of generations *9:* 1667
Alternative energy sources *1:* **111-118,** 114 (ill.), 115 (ill.), *6:* 1069
Alternative medicine *1:* **118-122**
Altimeter *2:* 266
Aluminum *1:* 122-124, 125 (ill.)
Aluminum family *1:* **122-126,** 125 (ill.)
Alzheimer, Alois *1:* 127
Alzheimer's disease *1:* 62, **126-130,** 128 (ill.)
Amazon basin *9:* 1774
American Red Cross *2:* 330
Ames test *2:* 408
Amino acid *1:* **130-131**
Aminoglycosides *1:* 158
Ammonia *7:* 1346
Ammonification *7:* 1343
Amniocentesis *2:* 322
Amoeba *1:* **131-134,** 132 (ill.)
Ampere *3:* 582, *4:* 737
Amère, André *4:* 737, *6:* 1212
Ampere's law *4:* 747
Amphibians *1:* **134-137,** 136 (ill.)
Amphiboles *1:* 191
Amphineura *7:* 1289
Amplitude modulation *8:* 1627
Amundsen, Roald *1:* 152
Anabolism *7:* 1255
Anaerobic respiration *9:* 1676
Anatomy *1:* **138-141,** 140 (ill.)
Anderson, Carl *1:* 163, *4:* 773
Andes Mountains *7:* 1301, *9:* 1775-1776
Andromeda galaxy *5:* 939 (ill.)
Anemia *1:* 8, *6:* 1220
Aneroid barometer *2:* 266
Anesthesia *1:* **142-145,** 143 (ill.)
Angel Falls *9:* 1774
Angiosperms *9:* 1729
Animal behavior *2:* 272
Animal hormones *6:* 1053
Animal husbandry *7:* 1433
Animals *1:* **145-147,** 146 (ill.), *6:* 1133-1134
Anorexia nervosa *4:* 712
Antarctic Treaty *1:* 153
Antarctica *1:* **147-153,** 148 (ill.), 152 (ill.)
Antennas *1:* **153-155,** 154 (ill.)
Anthrax *2:* 287
Antibiotics *1:* **155-159,** 157 (ill.)
Antibody and antigen *1:* **159-162,** *2:* 311
Anticyclones, cyclones and *3:* 608-610
Antidiuretic hormone *5:* 798
Antigens, antibodies and *1:* 159-162
Antimatter *1:* 163
Antimony *7:* 1348
Antiparticles *1:* **163-164**
Antiprotons *1:* 163
Antipsychotic drugs *8:* 1598
Antiseptics *1:* **164-166**
Anurans *1:* 136
Apennines *5:* 827
Apes *8:* 1572
Apgar Score *2:* 322
Aphasia *9:* 1798, 1799
Apollo 11 *9:* 1779, 1780 (ill.)
Apollo objects *1:* 202
Appalachian Mountains *7:* 1356
Appendicular skeleton *9:* 1741
Aquaculture *1:* **166-168,** 167 (ill.)
Arabian Peninsula. *See* **Middle East**
Arabic numbers. *See* **Hindu-Arabic number system**
Arachnids *1:* **168-171,** 170 (ill.)
Arachnoid *2:* 342
Arachnophobia *8:* 1497
Ararat, Mount *1:* 197

Archaeoastronomy *1:* **171-173,** 172 (ill.)
Archaeology *1:* **173-177,** 175 (ill.), 176 (ill.), *7:* 1323-1327
Archaeology, oceanic. *See* **Nautical archaeology**
Archaeopteryx lithographica 2: *312*
Archimedes *2:* 360
Archimedes' Principle *2:* 360
Argon *7:* 1349, 1350
Ariel *10:* 1954
Aristotle *1:* 138, *2:* 291, *5:* 1012, *6:* 1169
Arithmetic *1:* 97, **177-181,** *3:* 534-536
Arkwright, Edmund *6:* 1098
Armstrong, Neil *9:* 1779
Arnold of Villanova *2:* 404
ARPANET *6:* 1124
Arrhenius, Svante *1:* 14, *8:* 1495
Arsenic *7:* 1348
Arthritis *1:* **181-183,** 182 (ill.)
Arthropods *1:* **183-186,** 184 (ill.)
Artificial blood *2:* 330
Artificial fibers *1:* **186-188,** 187 (ill.)
Artificial intelligence *1:* **188-190,** *2:* 244
Asbestos *1:* **191-194,** 192 (ill.), *6:* 1092
Ascorbic acid. *See* **Vitamin C**
Asexual reproduction *9:* 1664 (ill.), 1665
Asia *1:* **194-200,** 195 (ill.), 198 (ill.)
Aspirin. *See* **Acetylsalicylic acid**
Assembly language *3:* 551
Assembly line *7:* 1238
Astatine *6:* 1035
Asterisms *3:* 560
Asteroid belt *1:* 201
Asteroids *1:* **200-204,** 203 (ill.), *9:* 1764
Asthenosphere *8:* 1535, 1536
Asthma *1:* **204-207,** 206 (ill.), *9:* 1681
Aston, William *7:* 1240
Astronomia nova *3:* 425
Astronomy, infrared *6:* 1100-1103
Astronomy, ultraviolet *10:* 1943-1946
Astronomy, x-ray *10:* 2038-2041
Astrophysics *1:* **207-209,** 208 (ill.)
Atherosclerosis *3:* 484
Atmosphere observation *2:* **215-217,** 216 (ill.)
Atmosphere, composition and structure *2:* **211-215,** 214 (ill.)
Atmospheric circulation *2:* **218-221,** 220 (ill.)
Atmospheric optical effects *2:* **221-225,** 223 (ill.)
Atmospheric pressure *2:* **225,** 265, *8:* 1571
Atom *2:* **226-229,** 227 (ill.)
Atomic bomb *7:* 1364, 1381
Atomic clocks *10:* 1895-1896
Atomic mass *2:* 228, **229-232**
Atomic number *4:* 777
Atomic theory *2:* **232-236,** 234 (ill.)
ATP *7:* 1258
Attention-deficit hyperactivity disorder (ADHD) *2:* **237-238**
Audiocassettes. *See* **Magnetic recording/audiocassettes**
Auer metal *6:* 1165
Auroras *2:* 223, 223 (ill.)
Australia *2:* **238-242,** 239 (ill.), 241 (ill.)
Australopithecus afarensis *6:* 1056, 1057 (ill.)
Australopithecus africanus *6:* 1056
Autistic savants. *See* **Savants**
Autoimmune diseases *1:* 162
Automation *2:* **242-245,** 244 (ill.)
Automobiles *2:* **245-251,** 246 (ill.), 249 (ill.)
Autosomal dominant disorders *5:* 966
Auxins *6:* 1051
Avogadro, Amadeo *7:* 1282
Avogadro's number *7:* 1282
Axial skeleton *9:* 1740
Axioms *1:* 179
Axle *6:* 1207
Ayers Rock *2:* 240
AZT *1:* 73

B

B-2 Stealth Bomber *1:* 78 (ill.)
Babbage, Charles *3:* 547
Babbitt, Seward *9:* 1691
Bacitracin *1:* 158
Bacteria *2:* **253-260,** 255 (ill.), 256 (ill.), 259 (ill.)
Bacteriophages *10:* 1974

Baekeland, Leo H. *8:* 1565
Bakelite *8:* 1565
Balard, Antoine *6:* 1034
Baldwin, Frank Stephen *2:* 371
Ballistics *2:* **260-261**
Balloons *1:* 75, *2:* ***261-265,*** *263 (ill.), 264 (ill.)*
Bardeen, John *10:* 1910
Barite *6:* 1093
Barium *1:* 105
Barnard, Christiaan *6:* 1043, *10:* 1926
Barometer *2:* **265-267,** 267 (ill.)
Barrier islands *3:* 500
Bases, acids and 1: 14-16
Basophils *2:* 329
Bats *4:* 721
Battery *2:* **268-270,** 268 (ill.)
Battle fatigue *9:* 1826
Beaches, coasts and *3:* 498-500
Becquerel, Henri *8:* 1630
Bednorz, Georg *10:* 1851
Behavior (human and animal), study of. *See* **Psychology**
Behavior *2:* **270-273,** 271 (ill.), 272 (ill.)
Behaviorism (psychology) *8:* 1595
Bell Burnell, Jocelyn *7:* 1340
Bell, Alexander Graham *10:* 1867 (ill.)
Benthic zone *7:* 1415
Benz, Karl Friedrich *2:* 246 (ill.)
Berger, Hans *9:* 1745
Beriberi *6:* 1219, *10:* 1982
Bernoulli's principle *1:* 40, 42, *5:* 884
Beryllium *1:* 103
Berzelius, Jöns Jakob *2:* 230
Bessemer converter *7:* 1445, *10:* 1916
Bessemer, Henry *10:* 1916
Beta carotene *10:* 1984
Beta particles *8:* 1632
Bichat, Xavier *1:* 141
Big bang theory *2:* **273-276,** 274 (ill.), *4:* 780
Bigelow, Julian *3:* 606
Binary number system *7:* 1397
Binary stars *2:* **276-278,** 278 (ill.)
Binomial nomenclature *2:* 337
Biochemistry *2:* **279-280**
Biodegradable *2:* **280-281**
Biodiversity *2:* **281-283,** 282 (ill.)
Bioenergy *1:* 117, *2:* **284-287,** 284 (ill.)
Bioenergy fuels *2:* 286
Biofeedback *1:* 119
Biological warfare *2:* **287-290**
Biological Weapons Convention Treaty *2:* 290
Biology *2:* **290-293,** *7:* 1283-1285
Bioluminescence *6:* 1198
Biomass energy. *See* **Bioenergy**
Biomes *2:* **293-302,** 295 (ill.), 297 (ill.), 301 (ill.)
Biophysics *2:* **302-304**
Bioremediation *7:* 1423
Biosphere 2 Project *2:* 307-309
Biospheres *2:* **304-309,** 306 (ill.)
Biot, Jean-Baptiste *7:* 1262
Biotechnology *2:* **309-312,** 311 (ill.)
Bipolar disorder *4:* 633
Birds *2:* **312-315,** 314 (ill.)
Birth *2:* **315-319,** 317 (ill.), 318 (ill.)
Birth control. *See* **Contraception**
Birth defects *2:* **319-322,** 321 (ill.)
Bismuth *7:* 1349
Bjerknes, Jacob *1:* 80, *10:* 2022
Bjerknes, Vilhelm *1:* 80, *10:* 2022
Black Death *8:* 1520
Black dwarf *10:* 2028
Black holes *2:* **322-326,** 325 (ill.), *9:* 1654
Blanc, Mont *5:* 827
Bleuler, Eugen *9:* 1718
Blood *2:* **326-330,** 328 (ill.), 330, *3:* 483
Blood banks *2:* 330
Blood pressure *3:* 483
Blood supply *2:* **330-333**
Blood vessels *3:* 482
Blue stars *9:* 1802
Bode, Johann *1:* 201
Bode's Law *1:* 201
Bogs *10:* 2025
Bohr, Niels *2:* 235
Bones. *See* **Skeletal system**
Bones, study of diseases of or injuries to. *See* **Orthopedics**
Boole, George *2:* 333
Boolean algebra *2:* **333-334**
Bopp, Thomas *3:* 529
Borax *1:* 126, *6:* 1094
Boreal coniferous forests *2:* 294
Bores, Leo *8:* 1617
Boron *1:* 124-126

Boron compounds *6:* 1094
Bort, Léon Teisserenc de *10:* 2021
Bosons *10:* 1831
Botany *2:* **334-337,** 336 (ill.)
Botulism *2:* 258, 288
Boundary layer effects *5:* 885
Bovine growth hormone *7:* 1434
Boyle, Robert *4:* 780
Boyle's law *5:* 960
Braham, R. R. *10:* 2022
Brahe, Tycho *3:* 574
Brain *2:* **337-351,** 339 (ill.), 341 (ill.)
Brain disorders *2:* 345
Brass *10:* 1920
Brattain, Walter *10:* 1910
Breathing *9:* 1680
Brewing *2:* **352-354,** 352 (ill.)
Bridges *2:* **354-358,** 357 (ill.)
Bright nebulae *7:* 1328
British system of measurement *10:* 1948
Bromine *6:* 1034
Bronchitis *9:* 1681
Bronchodilators *1:* 205
Brønsted, J. N. *1:* 15
Brønsted, J. N. *1:* 15
Bronze *2:* 401
Bronze Age *6:* 1036
Brown algae *1:* 95
Brown dwarf *2:* **358-359**
Brucellosis *2:* 288
Bryan, Kirk *8:* 1457
Bubonic plague *8:* 1518
Buckminsterfullerene *2:* 398, 399 (ill.)
Bugs. *See* **Insects**
Bulimia *4:* 1714-1716
Buoyancy *1:* 74, *2:* **360-361,** 360 (ill.)
Burial mounds *7:* 1298
Burns *2:* **361-364,** 362 (ill.)
Bushnell, David *10:* 1834
Butterflies *2:* **364-367,** 364 (ill.)
Byers, Horace *10:* 2022

C

C-12 *2:* 231
C-14 *1:* 176, *4:* 617
Cable television *10:* 1877
Cactus *4:* 635 (ill.)
CAD/CAM *2:* **369-370,** 369 (ill.)
Caffeine *1:* 34
Caisson *2:* 356
Calcite *3:* 422
Calcium *1:* 104 (ill.), 105
Calcium carbonate *1:* 104 (ill.)
Calculators *2:* **370-371,** 370 (ill.)
Calculus *2:* **371-372**
Calderas *6:* 1161
Calendars *2:* **372-375,** 374 (ill.)
Callisto *6:* 1148, 1149
Calories *2:* **375-376,** *6:* 1045
Calving (icebergs) *6:* 1078, 1079 (ill.)
Cambium *10:* 1927
Cambrian period *8:* 1461
Cameroon, Mount *1:* 51
Canadian Shield *7:* 1355
Canals *2:* **376-379,** 378 (ill.)
Cancer *2:* **379-382,** 379 (ill.), 381 (ill.), *10:* 1935
Canines *2:* **382-387,** 383 (ill.), 385 (ill.)
Cannabis sativa *6:* 1224, 1226 (ill.)
Cannon, W. B. *8:* 1516
Capacitor *4:* 749
Carbohydrates *2:* **387-389,** *7:* 1400
Carbon *2:* 396
Carbon compounds, study of. *See* **Organic chemistry**
Carbon cycle *2:* **389-393,** 391 (ill.)
Carbon dioxide *2:* **393-395,** 394 (ill.)
Carbon family *2:* **395-403,** 396 (ill.), 397 (ill.), 399 (ill.)
Carbon monoxide *2:* **403-406**
Carbon-12 *2:* 231
Carbon-14 *4:* 617
Carboniferous period *8:* 1462
Carborundum *1:* 2
Carcinogens *2:* **406-408**
Carcinomas *2:* 381
Cardano, Girolamo *8:* 1576
Cardiac muscle *7:* 1312
Cardiovascular system *3:* 480
Caries *4:* 628
Carlson, Chester *8:* 1502, 1501 (ill.)
Carnot, Nicholas *6:* 1118
Carnot, Sadi *10:* 1885
Carothers, Wallace *1:* 186
Carpal tunnel syndrome *2:* **408-410**
Cartography *2:* **410-412,** 411 (ill.)
Cascade Mountains *7:* 1358
Caspian Sea *5:* 823, 824

Cassini division *9:* 1711
Cassini, Giovanni Domenico *9:* 1711
Cassini orbiter *9:* 1712
CAT scans *2:* 304, *8:* 1640
Catabolism *7:* 1255
Catalysts and catalysis *3:* **413-415**
Catastrophism *3:* **415**
Cathode *3:* **415-416**
Cathode-ray tube *3:* **417-420,** 418 (ill.)
Cats. *See* **Felines**
Caucasus Mountains *5:* 823
Cavendish, Henry *6:* 1069, *7:* 1345
Caves *3:* **420-423,** 422 (ill.)
Cavities (dental) *4:* 628
Cayley, George *1:* 77
CDC *6:* 1180
CDs. *See* **Compact disc**
Celestial mechanics *3:* **423-428,** 427 (ill.)
Cell wall (plants) *3:* 436
Cells *3:* **428-436,** 432 (ill.), 435 (ill.)
Cells, electrochemical *3:* **436-439**
Cellular metabolism *7:* 1258
Cellular/digital technology *3:* **439-441**
Cellulose *2:* 389, *3:* **442-445,** 442 (ill.)
Celsius temperature scale *10:* 1882
Celsius, Anders *10:* 1882
Cenozoic era *5:* 990, *8:* 1462
Center for Disease Control (CDC) *6:* 1180
Central Asia *1:* 198
Central Dogma *7:* 1283
Central Lowlands (North America) *7:* 1356
Centrifuge *3:* **445-446,** 446 (ill.)
Cephalopoda *7:* 1289
Cephalosporin *1:* 158
Cepheid variables *10:* 1964
Ceramic *3:* **447-448**
Cerebellum *2:* 345
Cerebral cortex *2:* 343
Cerebrum *2:* 343
Čerenkov effect *6:* 1189
Cerium *6:* 1163
Cesium *1:* 102
Cetaceans *3:* **448-451,** 450 (ill.), *4:* 681 (ill.), *7:* 1416 (ill.)
CFCs *6:* 1032, *7:* 1453-1454, *8:* 1555,
Chadwick, James *2:* 235, *7:* 1338
Chain, Ernst *1:* 157
Chamberlain, Owen *1:* 163
Chancroid *9:* 1735, 1736
Chandra X-ray Observatory *10:* 2040
Chandrasekhar, Subrahmanyan *10:* 1854
Chandrasekhar's limit *10:* 1854
Chao Phraya River *1:* 200
Chaos theory *3:* **451-453**
Chaparral *2:* 296
Chappe, Claude *10:* 1864
Chappe, Ignace *10:* 1864
Charles's law *5:* 961
Charon *8:* 1541, 1541 (ill.), 1542
Chassis *2:* 250
Cheetahs *5:* 861
Chemical bond *3:* **453-457**
Chemical compounds *3:* 541-546
Chemical elements *4:* 774-781
Chemical equations *5 : 815-817*
Chemical equilibrium *5:* 817-820
Chemical warfare *3:* **457-463,** 459 (ill.), 461 (ill.)*6:* 1032
Chemiluminescence *6:* 1198
Chemistry *3:* **463-469,** 465 (ill.) ,467 (ill.), *8:* 1603
Chemoreceptors *8:* 1484
Chemosynthesis *7:* 1418
Chemotherapy *2:* 382
Chichén Itzá *1:* 173
Chicxulub *1:* 202
Childbed fever *1:* 164
Chimpanzees *8:* 1572
Chiropractic *1:* 120
Chladni, Ernst *1:* 17
Chlamydia *9:* 1735, 1736
Chlorination *6:* 1033
Chlorine *6:* 1032
Chlorofluorocarbons. *See* **CFCs**
Chloroform *1:* 142, 143, 143 (ill.)
Chlorophyll *1:* 103
Chlorophyta *1:* 94
Chloroplasts *3:* 436, *8:* 1506 (ill.)
Chlorpromazine *10:* 1906
Cholesterol *3:* **469-471,** 471 (ill.), *6:* 1042
Chorionic villus sampling *2:* 322, *4:* 790
Chromatic aberration *10:* 1871
Chromatography *8:* 1604
Chromosomes *3:* **472-476,** 472 (ill.), 475(ill.)

Chromosphere *10:* 1846
Chrysalis *2:* 366, *7:* 1261 (ill.)
Chrysophyta *1:* 93
Chu, Paul Ching-Wu *10:* 1851
Cigarette smoke *3:* **476-478,** 477 (ill.)
Cigarettes, addiction to *1:* 34
Ciliophora *8:* 1592
Circle *3:* **478-480,** 479 (ill.)
Circular acceleration *1:* 5
Circular accelerators *8:* 1479
Circulatory system *3:* **480-484,** 482 (ill.)
Classical conditioning *9:* 1657
Clausius, Rudolf *10:* 1885
Claustrophobia *8:* 1497
Climax community *10:* 1839
Clones and cloning *3:* **484-490,** 486 (ill.), 489 (ill.)
Clostridium botulinum *2:* 258
Clostridium tetani *2:* 258
Clouds *3:* **490-492,** 491 (ill.)
Coal *3:* **492-498,** 496 (ill.)
Coast and beach *3:* **498-500.** 500 (ill.)
Coastal Plain (North America) *7:* 1356
Cobalt-60 *7:* 1373
COBE (Cosmic Background Explorer) *2:* 276
COBOL *3:* 551
Cocaine *1:* 34, *3:* **501-505,** 503 (ill.)
Cockroaches *3:* **505-508,** 507 (ill.)
Coelacanth *3:* **508-511,** 510 (ill.)
Cognition *3:* **511-515,** 513 (ill.), 514 (ill.)
Cold fronts *1:* 81, 81 (ill.)
Cold fusion *7:* 1371
Cold-deciduous forests *5:* 909
Collins, Francis *6:* 1064
Collins, Michael *9:* 1779
Colloids *3:* **515-517,** 517 (ill.)
Color *3:* **518-522,** 521 (ill.)
Color blindness *5:* 971
Colorant *4:* 686
Colt, Samuel *7:* 1237
Columbus, Christopher *1:* 63
Coma *2:* 345
Combined gas law *5:* 960
Combustion *3:* **522-527,** 524 (ill.), *7:* 1441
Comet Hale-Bopp *3:* 529
Comet Shoemaker-Levy 9 *6:* 1151
Comet, Halley's *3:* 528
Comets *3:* **527-531,** 529 (ill.), *6:* 1151, *9:* 1765
Common cold *10:* 1978
Compact disc *3:* **531-533,** 532 (ill.)
Comparative genomics *6:* 1067
Complex numbers *3:* **534-536,** 534 (ill.), *6:* 1082
Composite materials *3:* **536-539**
Composting *3:* **539-541,** 539 (ill.)
Compound, chemical *3:* **541-546,** 543 (ill.)
Compton Gamma Ray Observatory *5:* 949
Compulsion *7:* 1405
Computer Aided Design and Manufacturing. *See* **CAD/CAM**
Computer languages *1:* 189, *3:* 551
Computer software *3:* **549-554,** 553 (ill.)
Computer, analog *3:* **546-547**
Computer, digital *3:* **547-549,** 548 (ill.)
Computerized axial tomography. *See* **CAT scans**
Concave lenses *6:* 1185
Conditioning *9:* 1657
Condom *3:* 563
Conduction *6:* 1044
Conductivity, electrical. *See* **Electrical conductivity**
Conservation laws *3:* **554-558.** 557 (ill.)
Conservation of electric charge *3:* 556
Conservation of momentum *7:* 1290
Conservation of parity *3:* 558
Constellations *3:* **558-560,** 559 (ill.)
Contact lines *5:* 987
Continental Divide *7:* 1357
Continental drift *8:* 1534
Continental margin *3:* **560-562**
Continental rise *3:* 562
Continental shelf *2:* 300
Continental slope *3:* 561
Contraception *3:* **562-566,** 564 (ill.)
Convection *6:* 1044
Convention on International Trade in Endangered Species *5:* 795
Convex lenses *6:* 1185
Cooke, William Fothergill *10:* 1865
Coordination compounds *3:* 544
Copernican system *3:* 574
Copper *10:* 1919-1921, 1920 (ill.)

Coral *3:* **566-569,** 567 (ill.), 568 (ill.)
Coral reefs *2:* 301
Core *4:* 711
Coriolis effect *2:* 219, *10:* 2029
Corona *10:* 1846
Coronary artery disease *6:* 1042
Coronas *2:* 225
Correlation *3:* **569-571**
Corson, D. R. *6:* 1035
Corti, Alfonso Giacomo Gaspare *4:* 695
Corticosteroids *1:* 206
Corundum *6:* 1094
Cosmetic plastic surgery *8:* 1530
Cosmic Background Explorer (COBE) *2:* 276
Cosmic dust *6:* 1130
Cosmic microwave background *2:* 275, *8:* 1637
Cosmic rays *3:* **571-573,** 573 (ill.)
Cosmology *1:* 171, *3:* **574-577**
Cotton *3:* **577-579,** 578 (ill.)
Coulomb *3:* **579-582**
Coulomb, Charles*3:* 579, *6:* 1212
Coulomb's law *4:* 744
Courtois, Bernard *6:* 1035
Courtship behaviors *2:* 273
Covalent bonding *3:* 455
Cowan, Clyde *10:* 1833
Coxwell, Henry Tracey *2:* 263
Coyotes *2:* 385
Craniotomy *8:* 1528
Creationism *3:* 577
Crick, Francis *3:* 473, *4:* 786, *5:* 973, 980 (ill.), 982, *7:* 1389
Cro-Magnon man *6:* 1059
Crop rotation *3:* 589
Crops *3:* **582-590,** 583 (ill.), 589 (ill.)
Crude oil *8:* 1492
Crust *4:* 709
Crustaceans *3:* **590-593,** 592 (ill.)
Cryobiology *3:* **593-595**
Cryogenics *3:* **595-601,** 597 (ill.)
Crystal *3:* **601-604,** 602 (ill.), 603 (ill.)
Curie, Marie *7:* 1450
Current electricity *4:* 742
Currents, ocean *3:* **604-605**
Cybernetics *3:* **605-608,** 607 (ill.)
Cyclamate *3:* **608**
Cyclone and anticyclone *3:* **608-610,** 609 (ill.)
Cyclotron *1:* 163, *8:* 1479, 1480 (ill.)
Cystic fibrosis *2:* 320
Cytokinin *6:* 1052
Cytoskeleton *3:* 434

D

Da Vinci, Leonardo *2:* 291, *4:* 691, *10:* 2020
Daddy longlegs *1:* 171
Dalton, John *2:* 226, 229, *2:* 232
Dalton's theory *2:* 232
Dam *4:* **611-613,** 612 (ill.)
Damselfly *1:* 184 (ill.)
Danube River *5:* 824
Dark matter *4:* **613-616,** 615 (ill.)
Dark nebulae *6:* 1131, *7:* 1330
Dart, Raymond *6:* 1056
Darwin, Charles *1:* 29, *6:* 1051, *8:* 1510
Dating techniques *4:* **616-619,** 618 (ill.)
Davy, Humphry *1:* 142, chlorine *6:* 1032, 1087
DDT (dichlorodiphenyltrichloroethane) *1:* 69, *4:* **619-622,** 620 (ill.)
De Bort, Léon Philippe Teisserenc *2:* 263
De Candolle, Augustin Pyrame *8:* 1509
De curatorum chirurgia *8:* 1528
De Forest, Lee *10:* 1961
De materia medica *5:* 877
De Soto, Hernando *7:* 1299
Dead Sea *1:* 196
Death *1:* 59-62
Decay *7:* 1442
Decimal system *1:* 178
Decomposition *2:* 392, *9:* 1648
Deimos *6:* 1229
Dementia *4:* **622-624,** 623 (ill.)
Democritus *2:* 226, 232
Dendrochronology. *See* **Tree-ring dating**
Denitrification *7:* 1343
Density *4:* **624-626,** 625 (ill.)
Dentistry *4:* **626-630,** 628 (ill.), 629 (ill.)
Depression *4:* **630-634,** 632 (ill.)
Depth perception *8:* 1483 (ill.), 1484

Dermis *6:* 1111
Desalination *7:* 1439, *10:* 2012
Descartes, René *6:* 1184
The Descent of Man *6:* 1055
Desert *2:* 296, *4:* **634-638,** 635 (ill.), 636 (ill.)
Detergents, soaps and *9:* 1756-1758
Devonian period *8:* 1461
Dew point *3:* 490
Dexedrine *2:* 238
Diabetes mellitus *4:* **638-640**
Diagnosis *4:* **640-644,** 643 (ill.)
Dialysis *4:* **644-646,** *7:* 1439
Diamond *2:* 396 (ill.), 397
Diencephalon *2:* 342
Diesel engine *4:* **646-647,** 647 (ill.)
Diesel, Rudolf *4:* 646, *10:* 1835
Differential calculus *2:* 372
Diffraction *4:* **648-651,** 648 (ill.)
Diffraction gratings *4:* 650
Diffusion *4:* **651-653,** 652 (ill.)
Digestion *7:* 1255
Digestive system *4:* **653-658,** 657 (ill.)
Digital audio tape *6:* 1211
Digital technology. *See* **Cellular/digital technology**
Dingoes *2:* 385, 385 (ill.)
Dinosaurs *4:* **658-665,** 660 (ill.), 663 (ill.), 664 (ill.)
Diodes *4:* **665-666,** *6:* 1176-1179
Dioscorides *5:* 878
Dioxin *4:* **667-669**
Dirac, Paul *1:* 163, *4:* 772
Dirac's hypothesis *1:* 163
Direct current (DC) *4:* 741
Dirigible *1:* 75
Disaccharides *2:* 388
Disassociation *7:* 1305
Disease *4:* **669-675,** 670 (ill.), 673 (ill.), *8:* 1518
Dissection *10:* 1989
Distillation *4:* **675-677,** 676 (ill.)
DNA *1:* 61, *2:* 310, *3:* 434, 473-474, *5:* 972-975, 980 (ill.), 981-984, *7:* 1389-1390
  forensic science *5:* 900
  human genome project *6:* 1060-1068
  mutation *7:* 1314-1316
Döbereiner, Johann Wolfgang *8:* 1486
Dogs. *See* **Canines**
Dollard, John *10:* 1871
Dolly (clone) *3:* 486
Dolphins *3:* 448, 449 (ill.)
Domagk, Gerhard *1:* 156
Domain names (computers) *6:* 1127
Dopamine *9:* 1720
Doppler effect *4:* **677-680,** 679 (ill.), *9:* 1654
Doppler radar *2:* 220 (ill.), *10:* 2023
Doppler, Christian Johann *9:* 1654
Down syndrome *2:* 319
Down, John Langdon Haydon *9:* 1713
Drake, Edwin L. *7:* 1419
Drebbel, Cornelius *10:* 1834
Drew, Richard *1:* 39
Drift nets *4:* **680-682,** 681 (ill.)
Drinker, Philip *8:* 1548
Drought *4:* **682-684,** 683 (ill.)
Dry cell (battery) *2:* 269
Dry ice *2:* 395
Drying (food preservation) *5:* 890
Dubois, Marie-Eugene *6:* 1058
Duodenum *4:* 655
Dura mater *2:* 342
Dust Bowl *4:* 682
Dust devils *10:* 1902
Dust mites *1:* 107, 108 (ill.)
DVD technology *4:* **684-686**
Dyes and pigments *4:* **686-690,** 688 (ill.)
Dynamite *5:* 845
Dysarthria *9:* 1798
Dyslexia *4:* **690-691,** 690 (ill.)
Dysphonia *9:* 1798
Dysprosium *6:* 1163

E

$E = mc^2$ *7:* 1363, 1366, *9:* 1662
Ear *4:* **693-698,** 696 (ill.)
Earth (planet) *4:* **698-702,** 699 (ill.)
Earth science *4:* **707-708**
Earth Summit *5:* 796
Earth's interior *4:* **708-711,** 710 (ill.)
Earthquake *4:* **702-707,** 705 (ill.), 706 (ill.)
Eating disorders *4:* **711-717,** 713 (ill.)
Ebola virus *4:* **717-720,** 719 (ill.)
Echolocation *4:* **720-722**
Eclipse *4:* **723-725,** 723 (ill.)

Ecological pyramid *5:* 894 (ill.), 896
Ecological system. *See* **Ecosystem**
Ecologists *4:* 728
Ecology *4:* **725-728**
Ecosystem *4:* **728-730,** 729 (ill.)
Edison, Thomas Alva *6:* 1088
EEG (electroencephalogram) *2:* 348, *9:* 1746
Eijkman, Christian *10:* 1981
Einstein, Albert *4:* 691, *7:* 1428, *9:* 1659 (ill.)
photoelectric effect *6:* 1188, *8:* 1504
space-time continuum *9:* 1777
theory of relativity *9:* 1659-1664
Einthoven, William *4:* 751
EKG (electrocardiogram) *4:* 751-755
El Niño *4:* **782-785,** 784 (ill.)
Elasticity *4:* **730-731**
Elbert, Mount *7:* 1357
Elbrus, Mount *5:* 823
Electric arc *4:* **734-737,** 735 (ill.)
Electric charge *4:* 743
Electric circuits *4:* 739, 740 (ill.)
Electric current *4:* 731, 734, **737-741,** 740 (ill.), 746, 748, 761, 767, 771, 773
Electric fields *4:* 743, 759
Electric motor *4:* **747-750,** 747 (ill.)
Electrical conductivity *4:* **731-734,** 735
Electrical force *3:* 579, 581-582, *4:* 744
Electrical resistance *4:* 732, 738, 746
Electricity *4:* **741-747,** 745 (ill.)
Electrocardiogram *4:* **751-755,** 753 (ill.), 754 (ill.)
Electrochemical cells *3:* 416, 436-439
Electrodialysis *4:* 646
Electroluminescence *6:* 1198
Electrolysis *4:* **755-758**
Electrolyte *4:* 755
Electrolytic cell *3:* 438
Electromagnet *6:* 1215
Electromagnetic field *4:* **758-760**
Electromagnetic induction *4:* **760-763,** 762 (ill.)
Electromagnetic radiation *8:* 1619
Electromagnetic spectrum *4:* **763-765,** *4:* 768, *6:* 1100, 1185, *8:* 1633, *9:* 1795
Electromagnetic waves *7:* 1268
Electromagnetism *4:* **766-768,** 766 (ill.)
Electron *4:* **768-773**
Electron gun *3:* 417
Electronegativity *3:* 455
Electronics *4:* **773-774,** 773 (ill.)
Electrons *4:* **768-773,** *10:* 1832, 1833
Electroplating *4:* 758
Element, chemical *4:* **774-781,** 778 (ill.)
Elementary algebra *1:* 98
Elements *4:* 775, 777, *8:* 1490, *10:* 1913
Embryo and embryonic development *4:* **785-791,** 788 (ill.)
Embryology *4:* 786
Embryonic transfer *4:* 790-791
Emphysema *9:* 1681
Encke division *9:* 1711
Encke, Johann *9:* 1711
Endangered species *5:* **793-796,** 795 (ill.)
Endangered Species Act *5:* 795
Endocrine system *5:* **796-801,** 799 (ill.)
Endoplasmic reticulum *3:* 433
Energy *5:* **801-805**
Energy and mass *9:* 1662
Energy conservation *1:* 117
Energy, alternative sources of *1:* 111-118, *6:* 1069
Engels, Friedrich *6:* 1097
Engineering *5:* **805-807,** 806 (ill.)
Engines *2:* 246, *6:* 1117, 1143, *9:* 1817, *10:* 1835
English units of measurement. *See* **British system of measurement**
ENIAC *3:* 551
Entropy *10:* 1886
Environment
air pollution *8:* 1552, 1553
effect of aerosols on *1:* 47,48
effect of carbon dioxide on *8:* 1554
effect of use of fossil fuels on *2:* 285, *7:* 1454
impact of aquaculture on *1:* 168
industrial chemicals *8:* 1557
ozone depletion *8:* 1555
poisons and toxins *8:* 1546
tropical deforestation *9:* 1744
water pollution *8:* 1556
Environmental ethics *5:* **807-811,** 809 (ill.), 810 (ill.)

Enzyme *5:* **812-815,** 812 (ill.), 814 (ill.)
Eosinophils *2:* 329
Epidemics *4:* 671
Epidermis *2:* 362, *6:* 1110
Epilepsy *2:* 347-349
Equation, chemical *5:* **815-817**
Equilibrium, chemical *5:* **817-820**
Equinox *9:* 1728
Erasistratus *1:* 138
Erbium *6:* 1163
Erosion *3:* 498, *5:* **820-823,** 821 (ill.), *9:* 1762
Erythroblastosis fetalis *9:* 1685
Erythrocytes *2:* 327
Escherichia coli *2:* 258
Esophagitis *4:* 656
Estrogen *5:* 801, *8:* 1599, 1600
Estuaries *2:* 300
Ethanol *1:* 89-91
Ether *1:* 142, 143
Ethics *3:* 489, *5:* 807-811
Ethylene glycol *1:* 91
Euglenoids *1:* 92
Euglenophyta *1:* 92
Eukaryotes *3:* 429, 432-435
Europa *6:* 1148, 1149
Europe *5:* **823-828,** 825 (ill.), 827 (ill.)
Europium *6:* 1163
Eutrophication *1:* 96, *5:* **828-831,** 830 (ill.)
Evans, Oliver *7:* 1237, *9:* 1820
Evaporation *5:* **831-832**
Everest, Mount *1:* 194
Evergreen broadleaf forests *5:* 909
Evergreen tropical rain forest *2:* 298
Evolution *1:* 26, 51, *5:* **832-839**
Excretory system *5:* **839-842**
Exhaust system *2:* 247
Exoplanets. *See* **Extrasolar planets**
Exosphere *2:* 214
Expansion, thermal *5:* 842-843, ***10:*** **1883-1884**
Expert systems *1:* 188
Explosives *5:* **843-847**
Extrasolar planets *5:* **847-848,** 846 (ill.)
Extreme Ultraviolet Explorer *6:* 1123
Exxon *Valdez* 7: *1424, 1425 (ill.)*
Eye *5:* **848-853,** 851 (ill.)
Eye surgery *8:* 1615-1618

F

Fahrenheit temperature scale *10:* 1882
Fahrenheit, Gabriel Daniel *10:* 1882
Far East *1:* 199
Faraday, Michael *4:* 761, 767, *6:* 1212
Farming. *See* **Agriculture**
Farnsworth, Philo *10:* 1875
Farsightedness *5:* 851
Father of
- acoustics *1:* 17
- American psychiatry *9:* 1713
- genetics *5:* 982
- heavier-than-air craft *1:* 77
- lunar topography *7:* 1296
- medicine *2:* 348
- modern chemistry *3:* 465
- modern dentistry *4:* 627
- modern evolutionary theory *5:* 833
- modern plastic surgery *8:* 1529
- rigid airships *1:* 75
- thermochemistry *3:* 525

Fats *6:* 1191
Fauchard, Pierre *4:* 627
Fault *5:* **855,** 856 (ill.)
Fault lines *5:* 987
Fear, abnormal or irrational. *See* **Phobias**
Feldspar *6:* 1094
Felines *5:* **855-864,** 861, 862 (ill.)
Fermat, Pierre de *7:* 1393, *8:* 1576
Fermat's last theorem *7:* 1393
Fermentation *5:* **864-867,** *10:* 2043
Fermi, Enrico *7:* 1365
Ferrell, William *2:* 218
Fertilization *5:* **867-870,** 868 (ill.)
Fertilizers *1:* 66
Fetal alcohol syndrome *1:* 87
Fiber optics *5:* **870-872,** 871 (ill.)
Fillings (dental) *4:* 628
Filovirus *4:* 717
Filtration *5:* **872-875**
Fingerprinting *5:* 900
Fire algae *1:* 94
First law of motion *6:* 1170
First law of planetary motion *7:* 1426
First law of thermodynamics *10:* 1885
Fish *5:* **875-878,** 876 (ill.)
Fish farming *1:* 166
Fishes, age of *8:* 1461
FitzGerald, George Francis *9:* 1660

Flash lock *6:* 1193
Fleas *8:* 1474, 1474 (ill.)
Fleischmann, Martin *7:* 1371
Fleming, Alexander *1:* 156
Fleming, John Ambrose *10:* 1961
Florey, Howard *1:* 157
Flower *5:* **878-862,** 881 (ill.)
Flu. *See* **Influenza**
Fluid dynamics *5:* **882-886**
Flukes *8:* 1473
Fluorescence *6:* 1197
Fluorescent light *5:* **886-888,** 888 (ill.)
Fluoridation *5:* **889-890**
Fluoride *5:* 889
Fluorine *6:* 1031-1032
Fluorspar *6:* 1095
Fly shuttle *6:* 1097
Fold lines *5:* 987
Food irradiation *5:* 893
Food preservation *5:* **890-894**
Food pyramid *7:* 1402, 1402 (ill.)
Food web and food chain *5:* **894-898,** 896 (ill.)
Ford, Henry *2:* 249 (ill.), *7:* 1237-1238
Forensic science *5:* **898-901,** 899 (ill.), *6:* 1067
Forestry *5:* **901-907,** 905 (ill.), 906 (ill.)
Forests *2:* 294-295, *5:* **907-914,** 909 (ill.), 910 (ill.), 913 (ill.)
Formula, chemical *5:* **914-917**
FORTRAN *3:* 551
Fossil and fossilization *5:* **917-921,** 919 (ill.), 920 (ill.), *6:* 1055, *7:* 1326 (ill.), *8:* 1458
Fossil fuels *1:* 112, *2:* 284, 392, *7:* 1319
Fossils, study of. *See* **Paleontology**
Foxes *2:* 384
Fractals *5:* **921-923,** 922 (ill.)
Fractions, common *5:* **923-924**
Fracture zones *7:* 1410
Francium *1:* 102
Free radicals *1:* 61
Freezing point *3:* 490
Frequency *4:* 763, *5:* **925-926**
Frequency modulation *8:* 1628
Freshwater biomes *2:* 298
Freud, Sigmund *8:* 1593, 1594
Friction *5:* **926-927**
Frisch, Otto *7:* 1362
Fronts *1:* 80-82
Fry, Arthur *1:* 39
Fujita Tornado Scale *10:* 1902
Fujita, T. Theodore *10:* 1902
Fuller, R. Buckminster *2:* 398
Fulton, Robert *10:* 1835
Functions (mathematics) *5:* **927-930,** *8:* 1485
Functional groups *7:* 1430
Fungi *5:* **930-934,** 932 (ill.)
Fungicides *1:* 67
Funk, Casimir *10:* 1982
Fyodorov, Svyatoslav N. *8:* 1617

G

Gabor, Dennis *6:* 1049
Gadolinium *6:* 1163
Gagarin, Yury *9:* 1778
Gaia hypothesis *5:* **935-940**
Galactic clusters *9:* 1808
Galaxies, active *5:* 944
Galaxies *5:* **941-945,** 941 (ill.), 943 (ill.), 9: 1806-1808
Galen, Claudius *1:* 139
Galileo Galilei *1:* 4, *5:* 1012, *6: 1149, 1170, 1184,* 7: *1296,* 10: *1869*
Galileo probe *6:* 1149
Gall bladder *3:* 469, *4:* 653, 655
Galle, Johann *7:* 1330
Gallium *1:* 126
Gallo, Robert *10:* 1978
Gallstones *3:* 469
Galvani, Luigi *2:* 304, *4:* 751
Gambling *1:* 36
Game theory *5:* **945-949**
Gamma rays *4:* 765, *5:* **949-951,** *8:* 1632
Gamma-ray burst *5:* **952-955,** 952 (ill.), 954 (ill.)
Ganges Plain *1:* 197
Ganymede *6:* 1148, 1149
Garbage. *See* **Waste management**
Gardening. *See* **Horticulture**
Gas, natural *7:* 1319-1321
Gases, electrical conductivity in *4:* 735
Gases, liquefaction of *5:* **955-958**
Gases, properties of *5:* **959-962,** 959 (ill.)
Gasohol *1:* 91

Gastropoda *7:* 1288
Gauss, Carl Friedrich *6:* 1212
Gay-Lussac, Joseph Louis *2:* 262
Gay-Lussac's law *5:* 962
Geiger counter *8:* 1625
Gell-Mann, Murray *10:* 1829
Generators *5:* **962-966,** 964 (ill.)
Genes *7:* 1248
Genes, mapping. *See* **Human Genome Project**
Genetic disorders *5:* **966-973,** 968 (ill.), 968 (ill.)
Genetic engineering *2:* 310, *5:* **973-980,** 976 (ill.), 979 (ill.)
Genetic fingerprinting *5:* 900
Genetics *5:* **980-986,** 983 (ill.)
Geneva Protocol *2:* 289
Genital herpes *9:* 1735
Genital warts *9:* 1735, 1737
Geocentric theory *3:* 574
Geologic map *5:* **986-989,** 988 (ill.)
Geologic time *5:* **990-993**
Geologic time scale *5:* 988
Geology *5:* **993-994,** 944 (ill.)
Geometry *5:* **995-999**
Geothermal energy *1:* 116
Gerbert of Aurillac *7:* 1396
Geriatrics *5:* 999
Germ warfare. *See* **Biological warfare**
Germanium *2:* 401
Gerontology *5:* **999**
Gestalt psychology *8:* 1595
Gibberellin *6:* 1051
Gilbert, William *6:* 1212
Gillies, Harold Delf *8:* 1529
Gills *5:* 877
Glacier *5:* **1000-1003,** 1002 (ill.)
Glaisher, James *2:* 263
Glass *5:* **1004-1006,** 1004 (ill.)
Glenn, John *9:* 1779
Gliders *1:* 77
Global Biodiversity Strategy *2:* 283
Global climate *5:* **1006-1009**
Globular clusters *9:* 1802, 1808
Glucose *2:* 388
Gluons *10:* 1831
Glutamate *9:* 1720
Glycerol *1:* 91
Glycogen *2:* 389
Gobi Desert *1:* 199
Goddard, Robert H. *9:* 1695 (ill.)
Goiter *6:* 1220
Gold *8:* 1566-1569
Goldberger, Joseph *6:* 1219
Golden-brown algae *1:* 93
Golgi body *3:* 433
Gondwanaland *1:* 149
Gonorrhea *9:* 1735, 1736
Gorillas *8:* 1572
Gould, Stephen Jay *1:* 32
Graphs and graphing *5:* **1009-1011**
Grasslands *2:* 296
Gravitons *10:* 1831
Gravity and gravitation *5:* **1012-1016,** 1014 (ill.)
Gray, Elisha *10:* 1867
Great Barrier Reef *2:* 240
Great Dividing Range *2:* 240
Great Lakes *6:* 1159
Great Plains *4:* 682, *7:* 1356
Great Red Spot (Jupiter) *6:* 1149, 1150 (ill.)
Great Rift Valley *1:* 49, 51
Great White Spot (Saturn) *9:* 1709
Green algae *1:* 94
Green flashes *2:* 224
Greenhouse effect *2:* 285, 393, *5:* 1003, **1016-1022,** 1020 (ill.). *8:* 1554, *10:* 1965
Gregorian calendar *2:* 373, 375
Grissom, Virgil *9:* 1779
Growth hormone *5:* 797
Growth rings (trees) *4:* 619
Guiana Highlands *9:* 1772
Guided imagery *1:* 119
Gum disease *4:* 630
Guth, Alan *2:* 276
Gymnophions *1:* 137
Gymnosperms *9:* 1729
Gynecology *5:* **1022-1024,** 1022 (ill.)
Gyroscope *5:* **1024-1025,** 1024 (ill.)

H

H.M.S. *Challenger* 7: *1413*
Haber process *7:* 1346
Haber, Fritz *7:* 1346
Hadley, George *2:* 218
Hahn, Otto *7:* 1361
Hale, Alan *3:* 529
Hale-Bopp comet *3:* 529

Hales, Stephen *2:* 337
Half-life *6:* **1027**
Halite *6:* 1096
Hall, Charles M. *1:* 124, *4:* 757
Hall, Chester Moore *10:* 1871
Hall, John *7:* 1237
Halley, Edmond *7:* 1262, *10:* 2020
Halley's comet *3:* 528
Hallucinogens *6:* **1027-1030**
Haloes *2:* 224
Halogens *6:* **1030-1036**
Hand tools *6:* **1036-1037,** 1036 (ill.)
Hard water *9:* 1757
Hargreaves, James *6:* 1098
Harmonices mundi *3:* 425
Harmonics *5:* 925
Hart, William Aaron *7:* 1320
Harvestmen (spider) *1:* 171, 170 (ill.)
Harvey, William *1:* 139, *2:* 292
Hazardous waste *10:* 2006-2007, 2006 (ill.)
HDTV *10:* 1879
Heart *6:* **1037-1043,** 1041 (ill.), 1042 (ill.)
Heart attack *6:* 1043
Heart diseases *3:* 470, *6:* 1040
Heart transplants *10:* 1926
Heart, measure of electrical activity. *See* **Electrocardiogram**
Heartburn *4:* 656
Heat *6:* **1043-1046**
Heat transfer *6:* 1044
Heat, measurement of. *See* **Calorie**
Heisenberg, Werner *8:* 1609
Heliocentric theory *3:* 574
Helium *7:* 1349
Helminths *8:* 1471 (ill.)
Hemiptera *6:* 1105
Hemodialysis *5:* 841
Henbury Craters *2:* 240
Henry, Joseph *4:* 761, *10:* 1865
Herbal medicine *1:* 120
Herbicides *1:* 54-59
Herculaneum *5:* 828
Heredity *7:* 1246
Hermaphroditism *9:* 1667
Heroin *1:* 32, 34
Herophilus *1:* 138
Héroult, Paul *1:* 124
Herpes *9:* 1737
Herschel, John *2:* 277
Herschel, William *2:* 277, *10:* 1871, 1952
Hertz, Heinrich *6:* 1188, *8:* 1502, 1626
Hess, Henri *3:* 525
Hevelius, Johannes *7:* 1296
Hewish, Antony *7:* 1340
Hibernation *6:* **1046-1048,** 1047 (ill.)
Himalayan Mountains *1:* 194, 197, *7:* 1301
Hindbrain *2:* 340
Hindenburg *1:* 76
Hindu-Arabic number system *1:* 178, *7:* 1396, *10:* 2047
Hippocrates *2:* 348
Histamine *6:* 1085
Histology *1:* 141
Historical concepts *6:* 1186
HIV (human immunodeficiency virus) *1:* 70, 72 (ill.), *8:* 1583
Hodgkin's lymphoma *6:* 1201
Hoffmann, Felix *1:* 6
Hofstadter, Robert *7:* 1339
Hogg, Helen Sawyer *10:* 1964
Holistic medicine *1:* 120
Holland, John *10:* 1835
Hollerith, Herman *3:* 549
Holmium *6:* 1163
Holograms and holography *6:* **1048-1050,** 1049 (ill.)
Homeopathy *1:* 120
Homeostasis *8:* 1516, 1517
Homo erectus *6:* 1058
Homo ergaster *6:* 1058
Homo habilis *6:* 1058
Homo sapiens *6:* 1055, 1058-1059
Homo sapiens sapiens *6:* 1059
Hooke, Robert *1:* 140, *4:* 731
Hooke's law *4:* 731
Hopewell mounds *7:* 1301
Hopkins, Frederick G. *10:* 1982
Hopper, Grace *3:* 551
Hormones *6:* **1050-1053**
Horticulture *6:* **1053-1054,** 1053 (ill.)
HTTP *6:* 1128
Hubble Space Telescope *9:* 1808, *10:* 1873, 1872 (ill.)
Hubble, Edwin *2:* 275, *7:* 1328, *9:* 1655, 1810
Human evolution *6:* **1054-1060,** 1057 (ill.), 1059 (ill.)
Human Genome Project *6:* **1060-**

1068, 1062 (ill.), 1065 (ill.), 1066 (ill.)
Human-dominated biomes *2:* 302
Humanistic psychology *8:* 1596
Humason, Milton *9:* 1655
Hurricanes *3:* 610
Hutton, James *10:* 1947
Huygens, Christiaan *6:* 1187, *9:* 1711
Hybridization *2:* 310
Hydrocarbons *7:* 1430-1431
Hydrogen *6:* **1068-1071,** 1068 (ill.)
Hydrologic cycle *6:* **1071-1075,** 1072 (ill.), 1073 (ill.)
Hydropower *1:* 113
Hydrosphere *2:* 305
Hydrothermal vents *7:* 1418, 1417 (ill.)
Hygrometer *10:* 2020
Hypertension *3:* 484
Hypertext *6:* 1128
Hypnotherapy *1:* 119
Hypotenuse *10:* 1932
Hypothalamus *2:* 342, 343
Hypothesis *9:* 1723

I

Icarus *1:* 74
Ice ages *6:* **1075-1078,** 1077 (ill.)
Icebergs *6:* **1078-1081**, 1080 (ill.), 1081 (ill.)
Idiot savants. *See* **Savants**
IgE *1:* 109
Igneous rock *9:* 1702
Ileum *4:* 656
Imaginary numbers *6:* **1081-1082**
Immune system *1:* 108, *6:* **1082-1087**
Immunization *1:* 161, *10:* 1060-1960
Immunoglobulins *1:* 159
Imprinting *2:* 272
In vitro fertilization *4:* 791
Incandescent light *6:* **1087-1090,** 1089 (ill.)
Inclined plane *6:* 1207
Indian peninsula *1:* 197
Indicator species *6:* **1090-1092,** 1091 (ill.)
Indium *1:* 126
Induction *4:* 760
Industrial minerals *6:* **1092-1097**
Industrial Revolution *1:* 28, *3:* 523, *6:* 1193, **1097-1100,** *7:* 1236, *9:* 1817
automation *2:* 242
effect on agriculture *1:* 63
food preservation *5:* 892
Infantile paralysis. *See* **Poliomyelitis**
Infants, sudden death. *See* **Sudden infant death syndrome (SIDS)**
Inflationary theory *2:* 275, 276
Influenza *4:* 672, *6:* 1084, *10:* 1978, 1979-1981
Infrared Astronomical Satellite *9:* 1808
Infrared astronomy *6:* **1100-1103,** 1102 (ill.)
Infrared telescopes *6:* 1101
Ingestion *4:* 653
Inheritance, laws of. *See* **Mendelian laws of inheritance**
Insecticides *1:* 67
Insects *6:* **1103-1106,** 1104 (ill.)
Insomnia *9:* 1747
Insulin *3:* 474, *4:* 638
Integers *1:* 180
Integral calculus *2:* 372
Integrated circuits *6:* **1106-1109,** 1108 (ill.), 1109 (ill.)
Integumentary system *6:* **1109-1112,** 1111 (ill.)
Interference *6:* **1112-1114,** 1113 (ill.)
Interferometer *6:* 1115 (ill.), 1116
Interferometry *10:* 1874, *6:* **1114-1116,** 1115 (ill.), 1116 (ill.)
Interferon *6:* 1084
Internal-combustion engines *6:* **1117-1119,** 1118 (ill.)
International Space Station *9:* 1788
International System of Units *2:* 376
International Ultraviolet Explorer *10:* 1946, *6:* **1120-1123,** 1122 (ill.)
Internet *6:* **1123-1130,** 1127 (ill.)
Interstellar matter *6:* **1130-1133,** 1132 (ill.)
Invertebrates *6:* **1133-1134,** 1134 (ill.)
Invertebrates, age of *8:* 1461
Io *6:* 1148, 1149
Iodine *6:* 1035
Ionic bonding *3:* 455
Ionization *6:* **1135-1137**
Ionization energy *6:* 1135

Ions *4:* 733
Iron *10:* 1915-1918
Iron lung *8:* 1548 (ill.)
Iron manufacture *6:* 1098
Irrational numbers *1:* 180, 181
Isaacs, Alick *6:* 1084
Islands *3:* 500, *6:* **1137-1141,** 1139 (ill.)
Isotopes *6:* **1141-1142,** *7:* 1241
IUE. *See* **International Ultraviolet Explorer**

J

Jackals *2:* 385
Jacquet-Droz, Henri *9:* 1691
Jacquet-Droz, Pierre *9:* 1691
James, William *8:* 1594
Jansky, Karl *8:* 1635
Java man *6:* 1058
Jefferson, Thomas *7:* 1300
Jejunum *4:* 655
Jenner, Edward *1:* 161, *10:* 1957
Jet engines *6:* **1143-1146,** 1143 (ill.), 1145 (ill.)
Jet streams *2:* 221, *4:* 783, *7:* 1293
Jones, John *8:* 1529
Joule *6:* 1045
Joule, James *10:* 1885
Joule-Thomson effect *3:* 597
Jupiter (planet) *6:* **1146-1151,** 1147 (ill.), 1150 (ill.)

K

Kangaroos and wallabies *6:* **1153-1157,** 1155 (ill.)
Kant, Immanuel *9:* 1765
Kay, John *6:* 1097
Kelvin scale *10:* 1882
Kelvin, Lord. *See* **Thomson, William**
Kenyanthropus platyops *6:* 1056
Kepler, Johannes *3:* 425, 574, *7:* 1426
Keratin *6:* 1110
Kettlewell, Henry Bernard David *1:* 28
Kidney dialysis *4:* 645
Kidney stones *5:* 841
Kilimanjaro, Mount *5:* 1000
Kinetic theory of matter *7:* 1243
King, Charles G. *6:* 1219
Klein bottle *10:* 1899
Knowing. *See* **Cognition**
Klein, Felix *10:* 1899
Koch, Robert *2:* 292
Köhler, Wolfgang *8:* 1595
Kraepelin, Emil *9:* 1718
Krakatoa *10:* 1998
Krypton *7:* 1349, 1352
Kuiper Disk *3:* 530
Kuiper, Gerald *3:* 530, *7:* 1333
Kwashiorkor *6:* 1218, *7:* 1403

L

La Niña *4:* 782
Lacrimal gland *5:* 852
Lactose *2:* 388
Lager *2:* 354
Lake Baikal *1:* 198
Lake Huron *6:* 1162 (ill.)
Lake Ladoga *5:* 824
Lake Michigan *7:* 1354 (ill.)
Lake Superior *6:* 1159
Lake Titicaca *6:* 1159
Lakes *6:* **1159-1163,** 1161 (ill.), 1162 (ill.)
Lamarck, Jean-Baptiste *1:* 28
Lambert Glacier *1:* 149
Laminar flow *1:* 40
Landfills *10:* 2007, 2008 (ill.)
Language *3:* 515
Lanthanides *6:* **1163-1166**
Lanthanum *6:* 1163
Laplace, Pierre-Simon *2:* 323, *9:* 1765
Large intestine *4:* 656
Laryngitis *9:* 1681
Laser eye surgery *8:* 1617
Lasers *6:* **1166-1168,** 1168 (ill.)
LASIK surgery *8:* 1617
Laurentian Plateau *7:* 1355
Lava *10:* 1995
Lavoisier, Antoine Laurent *3:* 465, 524, *6:* 1069, *7:* 1444
Law of conservation of energy *3:* 555, *10:* 1885
Law of conservation of mass/matter *3:* 554, *5:* 816
Law of conservation of momentum *7:* 1290

Law of dominance *7:* 1249
Law of electrical force *3:* 579
Law of independent assortment *7:* 1249
Law of planetary motion *3:* 425
Law of segregation *7:* 1249
Law of universal gravitation *3:* 426, *7:* 1427
Lawrence, Ernest Orlando *8:* 1479
Laws of motion *3:* 426, *6:* **1169-1171,** *7:* 1235, 1426
Le Verrier, Urbain *7:* 1330
Lead *2:* 402-403
Leakey, Louis S. B. *6:* 1058
Learning disorders *4:* 690
Leaf *6:* **1172-1176,** 1172 (ill.), 1174 (ill.)
Leavitt, Henrietta Swan *10:* 1964
Leclanché, Georges *2:* 269
LED (light-emitting diode) *6:* **1176-1179,** 1177 (ill.), 1178 (ill.)
Leeuwenhoek, Anton van *2:* 253, *6:* 1184, *8:* 1469
Legionella pneumophilia *6:* 1181, 1182
Legionnaire's disease *6:* **1179-1184,** 1182 (ill.)
Leibniz, Gottfried Wilhelm *2:* 371, 372, *7:* 1242
Lemaître, Georges-Henri *3:* 576
Lemurs *8:* 1572
Lenoir, Jean-Joseph Éttien *6:* 1119
Lenses *6:* **1184-1185,** 1184 (ill.)
Lentic biome *2:* 298
Leonid meteors *7:* 1263
Leonov, Alexei *9:* 1779
Leopards *5:* 860
Leptons *10:* 1830
Leucippus *2:* 226
Leukemia *2:* 380
Leukocytes *2:* 328
Lever *6:* 1205, 1206 (ill.)
Levy, David *6:* 1151
Lewis, Gilbert Newton *1:* 15
Liber de Ludo Aleae *8:* 1576
Lice *8:* 1474
Life, origin of *4:* 702
Light *6:* 1087-1090, **1185-1190**
Light, speed of *6:* 1190
Light-year *6:* **1190-1191**
Lightning *10:* 1889, 1889 (ill.)
Limbic system *2:* 345
Liming agents *1:* 66
Lind, James *6:* 1218, *10:* 1981
Lindenmann, Jean *6:* 1084
Linear acceleration *1:* 4
Linear accelerators *8:* 1477
Linnaeus, Carolus *2:* 292, 337
Lions *5:* 860
Lipids *6:* **1191-1192,** *7:* 1400
Lippershey, Hans *10:* 1869
Liquid crystals *7:* 1244-1245
Liquor. *See* **Alcohol (liquor)**
Lister, Joseph *1:* 165
Lithium *1:* 100
Lithosphere *8:* 1535, 1536
Litmus test *8:* 1496
Locks (water) *6:* **1192-1195,** 1193 (ill.)
Logarithms *6:* **1195**
Long, Crawford W. *1:* 143
Longisquama insignis *2:* 312
Longitudinal wave *10:* 2015
Lord Kelvin. *See* **Thomson, William**
Lorises *8:* 1572
Lotic *2:* 299
Lotic biome *2:* 299
Lowell, Percival *8:* 1539
Lowry, Thomas *1:* 15
LSD *6:* 1029
Lucy (fossil) *6:* 1056, 1057 (ill.)
Luminescence *6:* **1196-1198,** 1196 (ill.)
Luna *7:* 1296
Lunar eclipses *4:* 725
Lunar Prospector *7:* 1297
Lung cancer *9:* 1682
Lungs *9:* 1679
Lunisolar calendar *2:* 374
Lutetium *6:* 1163
Lymph *6:* 1199
Lymph nodes *6:* 1200
Lymphatic system *6:* **1198-1202**
Lymphocytes *2:* 329, *6:* 1085, 1200 (ill.)
Lymphoma *2:* 380, *6:* 1201
Lysergic acid diethylamide. *See* **LSD**

M

Mach number *5:* 883
Mach, L. *6:* 1116
Machines, simple *6:* **1203-1209,** 1206 (ill.), 1208 (ill.)

Mackenzie, K. R. *6:* 1035
Magellan *10:* 1966
Magma *10:* 1995
Magnesium *1:* 103
Magnetic fields *4:* 759
Magnetic fields, stellar. *See* **Stellar magnetic fields**
Magnetic recording/audiocassette *6:* **1209-1212,** 1209 (ill.), 1211 (ill.)
Magnetic resonance imaging *2:* 304
Magnetism *6:* **1212-1215,** 1214 (ill.)
Malnutrition *6:* **1216-1222,** 1221 (ill.)
Malpighi, Marcello *1:* 139
Mammals *6:* **1222-1224,** 1223 (ill.)
Mammals, age of *8:* 1462
Mangrove forests *5:* 909
Manhattan Project *7:* 1365, 1380
Manic-depressive illness *4:* 631
Mantle *4:* 710
Manufacturing. *See* **Mass production**
MAP (Microwave Anisotroy Probe) *2:* 276
Maps and mapmaking. *See* **Cartography**
Marasmus *6:* 1218
Marconi, Guglielmo *8:* 1626
Marie-Davy, Edme Hippolyte *10:* 2021
Marijuana *6:* **1224-1227,** 1226 (ill.), 1227 (ill.)
Marine biomes *2:* 299
Mariner 10 *7:* 1250
Mars (planet) *6:* **1228-1234,** 1228 (ill.), 1231 (ill.), 1232 (ill.)
Mars Global Surveyor *6:* 1230
Mars Pathfinder *6:* 1232
Marshes *10:* 2025
Maslow, Abraham *8:* 1596
Mass *7:* **1235-1236**
Mass production *7:* **1236-1239,** 1238 (ill.)
Mass spectrometry *7:* **1239-1241,** 1240 (ill.), *8:* 1604
Mastigophora *8:* 1592
Mathematics *7:* **1241-1242**
- imaginary numbers *6:* 1081
- logarithms *6:* 1195
- multiplication *7:* 1307
- number theory *7:* 1393
- probability theory *8:* 1575
- proofs *8:* 1578
- statistics *9:* 1810
- symbolic logic *10:* 1859
- topology *10:* 1897-1899
- trigonometry *10:* 1931-1933
- zero *10:* 2047

Matter, states of *7:* **1243-1246,** 1243 (ill.)
Maxwell, James Clerk *4:* 760, 767, *6:* 1213, *8:* 1626
Maxwell's equations *4:* 760
McKay, Frederick *5:* 889
McKinley, Mount *7:* 1302
McMillan, Edwin *1:* 24
Measurement. *See* **Units and standards**
Mechanoreceptors *8:* 1484
Meditation *1:* 119
Medulla oblongata *2:* 340
Meiosis *9:* 1666
Meissner effect *10:* 1851 (ill.)
Meitner, Lise *7:* 1362
Mekong River *1:* 200
Melanin *6:* 1110
Melanomas *2:* 380
Memory *2:* 344, *3:* 515
Mendel, Gregor *2:* 337, *4:* 786, *7:* 1247
Mendeleev, Dmitry *4:* 777, *8:* 1487
Mendelian laws of inheritance *7:* **1246-1250,** 1248 (ill.)
Meninges *2:* 342
Menopause *1:* 59, *2:* 410, *5:* 800, 1020
Menstruation *1:* 59, *5:* 800, 1020, *8:* 1599
Mental illness, study and treatment of. *See* **Psychiatry**
Mercalli scale *4:* 704
Mercalli, Guiseppe *4:* 704
Mercury *10:* 1921-1923, 1922 (ill.)
Mercury (planet) *7:* **1250-1255,** 1251 (ill.), 1252 (ill.)
Mercury barometers *2:* 265
Méré, Chevalier de *8:* 1576
Mescaline *6:* 1029
Mesosphere *2:* 213
Mesozoic era *5:* 990, *8:* 1462
Metabolic disorders *7:* **1254-1255,** 1254 (ill.), 1257
Metabolism *7:* **1255-1259**
Metalloids *7:* 1348
Metamorphic rocks *9:* 1705
Metamorphosis *7:* **1259-1261**

Meteorograph *2:* 215
Meteors and meteorites *7:* **1262-1264**
Methanol *1:* 89
Metric system *7:* **1265-1268,** *10:* 1949
Mettauer, John Peter *8:* 1529
Meyer, Julius Lothar *4:* 777, *8:* 1487
Michell, John *2:* 323
Michelson, Albert A. *6:* 1114, *6:* 1187
Microwave Anisotropy Probe (MAP) *2:* 276
Microwave communication *7:* **1268-1271,** 1270 (ill.)
Microwaves *4:* 765
Mid-Atlantic Ridge *7:* 1303, 1409
Midbrain *2:* 340
Middle East *1:* 196
Mifepristone *3:* 565
Migraine *2:* 349, 350
Migration (animals) *7:* **1271-1273,** 1272 (ill.)
Millennium *2:* 375
Millikan, Robert Andrew *4:* 771
Minerals *6:* 1092-1097, *7:* 1401, **1273-1278,** 1276 (ill.), 1277 (ill.)
Mining *7:* **1278-1282,** 1281 (ill.)
Mir *9:* 1781
Mirages *2:* 222
Miranda *10:* 1954
Misch metal *6:* 1165
Missiles. *See* **Rockets and missiles**
Mission *9:* 1787
Mississippi River *7:* 1355
Mississippian earthern mounds *7:* 1301
Missouri River *7:* 1355
Mitchell, Mount *7:* 1356
Mites *1:* 170
Mitochondira *3:* 436
Mitosis *1:* 133, *9:* 1665
Mobile telephones *3:* 441
Möbius strip *10:* 1899
Möbius, Augustus Ferdinand *10:* 1899
Model T (automobile) *7:* 1237, 1238
Modulation *8:* 1627
Moho *4:* 709
Mohorovičiá discontinuity *4:* 709
Mohorovičiá, Andrija *4:* 709
Mohs scale *1:* 3
Mohs, Friedrich *1:* 3
Mole (measurement) *7:* **1282-1283**
Molecular biology *7:* **1283-1285**
Molecules *7:* **1285-1288,** 1286 (ill.)
Mollusks *7:* **1288-1290,** 1289 (ill.)
Momentum *7:* **1290-1291**
Monkeys *8:* 1572
Monoclonal antibodies *1:* 162, *2:* 311
Monocytes *2:* 329
Monoplacophora *7:* 1289
Monosaccharides *2:* 388
Monotremes *6:* 1224
Monsoons *7:* **1291-1294**
Mont Blanc *5:* 827
Montgolfier, Jacques *2:* 262
Montgolfier, Joseph *2:* 262
Moon *7:* **1294-1298,** 1295 (ill.), 1297 (ill.)
  affect on tides *10:* 1890
  Apollo 11 9: *1779*
Morley, Edward D. *6:* 1187
Morphine *1:* 32, 33
Morse code *10:* 1866
Morse, Samuel F. B. *10:* 1865
Morton, William *1:* 143
Mosquitoes *6:* 1106, *8:* 1473
Motion, planetary, laws of *7:* 1426
Motors, electric. *See* **Electric motors**
Mounds, earthen *7:* **1298-1301,** 1299 (ill.)
Mount Ararat *1:* 197
Mount Cameroon *1:* 51
Mount Elbert *7:* 1357
Mount Elbrus *5:* 823
Mount Everest *1:* 194
Mount Kilimanjaro *1:* 53 (ill.), *5:* 1003, *7:* 1303
Mount McKinley *7:* 1302, 1354
Mount Mitchell *7:* 1356
Mount Robson *7:* 1357
Mount St. Helens *10:* 1996, 1998 (ill.)
Mountains *7:* **1301-1305,** 1304 (ill.)
Movable bridges *2:* 358
mRNA *7:* 1285
Müller, Karl Alex *10:* 1851
Multiple personality disorder *7:* **1305-1307**
Multiplication *7:* **1307-1309**
Muscular dystrophy *7:* 1313, 1337
Muscular system *7:* **1309-1313,** 1311 (ill.), 1312 (ill.), 1313 (ill.)
Mushrooms *6:* 1028
Mutation *7:* **1314-1317,** 1316 (ill.)
Mysophobia *8:* 1497

N

Napier, John *6:* 1195
Narcolepsy *9:* 1748
Narcotics *1:* 32
Natural gas *7:* **1319-1321**
Natural language processing *1:* 189
Natural numbers *1:* 180, *7:* **1321-1322**
Natural selection *1:* 29-30, *2:* 292, *5:* 834, 837-839
Natural theology *1:* 27
Naturopathy *1:* 122
Nautical archaeology *7:* **1323-1327,** 1325 (ill.), 1326 (ill.)
Nautilus *10:* 1835
Navigation (animals) *7:* 1273
Neanderthal man *6:* 1059
Neap tides *10:* 1892
NEAR Shoemaker *1:* 203-204, *9:* 1787
Nearsightedness *5:* 851
Nebula *7:* **1327-1330,** 1329 (ill.)
Nebular hypothesis *9:* 1765
Negative reinforcement. *See* **Reinforcement, positive and negative**
Nematodes *8:* 1471
Neodymium *6:* 1163
Neolithic Age *1:* 62
Neon *7:* 1349, 1352
Neptune (planet) *7:* **1330-1333,** 1331 (ill.)
Nereid *7:* 1333
Nerve nets *7:* 1333
Nervous system *7:* **1333-1337,** 1335 (ill.)
Neurons *7:* 1333 (ill.)
Neurotransmitters *1:* 128, *2:* 350, *4:* 631, *7:* 1311, *9:* 1720
Neutralization *1:* 15
Neutrinos *10:* 1832, 1833
Neutron *2:* 235, *7:* **1337-1339,** *10:* 1830, 1832
Neutron stars *7:* **1339-1341,** 1341 (ill.)
Neutrophils *2:* 329
Newcomen, Thomas *9:* 1818
Newton, Isaac *1:* 4, *5:* 1012, *6:* 1184, *7:* 1242
  calculus *2:* 372
  corpsucular theory of light *6:* 1187
  laws of motion *6:* 1169-1171, *7:* 1427
  reflector telescope *10:* 1871
Ngorongoro Crater *1:* 52
Niacin. *See* **Vitamin B3**
Nicotine *1:* 34, *3:* 478
Night blindness *6:* 1220
Nile River *9:* 1686
Nipkow, Paul *10:* 1875
Nitrification *7:* 1343, 1344
Nitrogen *7:* 1344, 1345 (ill.)
Nitrogen cycle *7:* **1342-1344,** 1342 (ill.)
Nitrogen family *7:* **1344-1349,** 1345 (ill.)
Nitrogen fixation *7:* 1342
Nitroglycerin *5:* 844
Nitrous oxide *1:* 142
Nobel, Alfred *5:* 845
Noble gases *7:* **1349-1352,** 1350 (ill.)
Non-Hodgkin's lymphoma *6:* 1201
Nondestructive testing *10:* 2037
North America *7:* **1352-1358,** 1353 (ill.), 1354 (ill.), 1357 (ill.)
Novas *7:* **1359-1360,** 1359 (ill.)
NSFNET *6:* 1127
Nuclear fission *7:* **1361-1366,** 1365 (ill.), 1381
Nuclear fusion *7:* **1366-1371,** 1370 (ill.)
Nuclear medicine *7:* **1372-1374**
Nuclear Non-Proliferation Treaty *7:* 1387
Nuclear power *1:* 113, *7:* **1374-1381,** 1376 (ill.), 1379 (ill.), *10:* 1836
Nuclear power plant *7:* 1374, 1379 (ill.)
Nuclear reactor *7:* 1365
Nuclear waste management *7:* 1379
Nuclear weapons *7:* **1381-1387**
Nucleic acids *7:* **1387-1392,** 1391 (ill.), 1392 (ill.)
Nucleotides *3:* 473, *7:* 1388
Nucleus (cell) *3:* 434
Number theory *7:* 1322, **1393-1395**
Numbers, imaginary *6:* 1081
Numbers, natural *7:* 1321
Numeration systems *7:* **1395-1399**
Nutrient cycle *2:* 307
Nutrients *7:* 1399
Nutrition *6:* 1216, *7:* **1399-1403,** 1402 (ill.)
Nylon *1:* 186, *8:* 1533

O

Oberon *10:* 1954
Obesity *4:* 716
Obsession *7:* **1405-1407**
Obsessive-compulsive disorder *7:* 1405
Obsessive-compulsive personality disorder *7:* 1406
Occluded fronts *1:* 82
Ocean *7:* **1407-1411,** 1407 (ill.)
Ocean currents *3:* 604-605
Ocean ridges *7:* 1410 (ill.)
Ocean zones *7:* **1414-1418**
Oceanic archaeology. *See* **Nautical archaeology**
Oceanic ridges *7:* 1409
Oceanography *7:* **1411-1414,** 1412 (ill.), 1413 (ill.)
Octopus *7:* 1289
Oersted, Hans Christian *1:* 124, *4:* 760, 766, *6:* 1212
Offshore drilling *7:* 1421
Ohio River *7:* 1355
Ohm (O) *4:* 738
Ohm, Georg Simon *4:* 738
Ohm's law *4:* 740
Oil drilling *7:* **1418-1422,** 1420 (ill.)
Oil pollution *7:* 1424
Oil spills *7:* **1422-1426,** 1422 (ill.), 1425 (ill.)
Oils *6:* 1191
Olduvai Gorge *6:* 1058
Olfaction. *See* **Smell**
On the Origin of Species by Means of Natural Selection *6:* 1054
On the Structure of the Human Body *1:* 139
O'Neill, J. A. *6:* 1211
Onnes, Heike Kamerlingh *10:* 1850
Oort cloud *3:* 530
Oort, Jan *8:* 1637
Open clusters *9:* 1808
Open ocean biome *2:* 299
Operant conditioning *9:* 1658
Ophediophobia *8:* 1497
Opiates *1:* 32
Opium *1:* 32, 33
Orangutans *8:* 1572, 1574 (ill.)
Orbit *7:* **1426-1428**
Organ of Corti *4:* 695
Organic chemistry *7:* **1428-1431**
Organic families *7:* 1430
Organic farming *7:* **1431-1434,** 1433 (ill.)
Origin of life *4:* 702
Origins of algebra *1:* 97
Orizaba, Pico de *7:* 1359
Orthopedics *7:* **1434-1436**
Oscilloscopes *10:* 1962
Osmosis *4:* 652, *7:* **1436-1439,** 1437 (ill.)
Osmotic pressure *7:* 1436
Osteoarthritis *1:* 181
Osteoporosis *9:* 1742
Otitis media *4:* 697
Otosclerosis *4:* 697
Ovaries *5:* 800
Oxbow lakes *6:* 1160
Oxidation-reduction reactions *7:* **1439-1442,** *9:* 1648
Oxone layer *7:* 1452 (ill.)
Oxygen family *7:* **1442-1450,** 1448 (ill.)
Ozone *7:* **1450-1455,** 1452 (ill.)
Ozone depletion *1:* 48, *8:* 1555
Ozone layer *7:* 1451

P

Packet switching *6:* 1124
Pain *7:* 1336
Paleoecology *8:* **1457-1459,** 1458 (ill.)
Paleontology *8:* **1459-1462,** 1461 (ill.)
Paleozoic era *5:* 990, *8:* 1461
Paleozoology *8:* 1459
Panama Canal *6:* 1194
Pancreas *4:* 655, *5:* 798
Pangaea *8:* 1534, 1536 (ill.)
Pap test *5:* 1020
Papanicolaou, George *5:* 1020
Paper *8:* **1462-1467,** 1464 (ill.), 1465 (ill.), 1466 (ill.)
Papyrus *8:* 1463
Paracelsus, Philippus Aureolus *1:* 84
Parasites *8:* **1467-1475,** 1471 (ill.), 1472 (ill.), 1474 (ill.)
Parasitology *8:* 1469
Parathyroid glands *5:* 798
Paré, Ambroise *8:* 1580, *10:* 1855
Parkinson's disease *1:* 62
Parsons, Charles A. *9:* 1820
Particle accelerators *8:* **1475-1482,**

1478 (ill.), 1480 (ill.)
Particulate radiation *8:* 1620
Parturition. *See* **Birth**
Pascal, Blaise *2:* 370, *8:* 1576
Pascaline *2:* 371
Pasteur, Louis *1:* 161, 165, *2:* 292, *10:* 1958, 1990, 1959 (ill.)
Pavlov, Ivan P. *9:* 1657, *10:* 1990
Pelagic zone *7:* 1415
Pellagra *6:* 1219
Penicillin *1:* 155, 156, 157 (ill.)
Peppered Moth *1:* 28
Perception *8:* **1482-1485,** 1483 (ill.)
Periodic function *8:* **1485-1486,** 1485 (ill.)
Periodic table *4:* 777, 778 (ill.), *8:* 1489, **1486-1490,** 1489 (ill.), 1490 (ill.)
Periscopes *10:* 1835
Perrier, C. *4:* 775, *10:* 1913
Perseid meteors *7:* 1263
Persian Gulf War *3:* 462, *7:* 1425
Pesticides *1:* 67, 67 (ill.), 68 (ill.), *4:* 619-622
PET scans *2:* 304, *8:* 1640
Petroglyphs and pictographs *8:* **1491-1492,** 1491 (ill.)
Petroleum *7:* 1418, 1423, ***8:* 1492-1495**
Peyote *6:* 1029
Pfleumer, Fritz *6:* 1211
pH *8:* **1495-1497**
Phaeophyta *1:* 95
Phages *10:* 1974
Phenothiazine *10:* 1906
Phenylketonuria *7:* 1254
Phloem *6:* 1175, *8:* 1523
Phobias *8:* **1497-1498**
Phobos *6:* 1229
Phosphates *6:* 1095
Phosphorescence *6:* 1197
Phosphorus *7:* 1347
Photochemistry *8:* **1498-1499**
Photocopying *8:* **1499-1502,** 1500 (ill.), 1501 (ill.)
Photoelectric cell *8:* 1504
Photoelectric effect *6:* 1188, *8:* **1502-1505**
Photoelectric theory *8:* 1503
Photoreceptors *8:* 1484
Photosphere *10:* 1846
Photosynthesis *2:* 306, 388, 391, *8:* **1505-1507**
Phototropism *6:* 1051, *8:* **1508-1510,** 1508 (ill.)
Physical therapy *8:* **1511-1513,** 1511 (ill.)
Physics *8:* **1513-1516**
Physiology *8:* **1516-1518**
Phytoplankton *8:* 1520, 1521
Pia mater *2:* 342
Piazzi, Giuseppe *1:* 201
Pico de Orizaba *7:* 1359
Pictographs, petroglyphs and *8:* 1491-1492
Pigments, dyes and *4:* 686-690
Pineal gland *5:* 798
PKU (phenylketonuria) *7:* 1254
Place value *7:* 1397
Plages *10:* 1846
Plague *8:* **1518-1520,** 1519 (ill.)
Planck's constant *8:* 1504
Plane *6:* 1036 (ill.), 1207
Planetary motion, laws of *7:* 1426
Plankton *8:* **1520-1522**
Plant behavior *2:* 270
Plant hormones *6:* 1051
Plants *1:* 91, *2:* 337, 388, 392, *8:* 1505, **1522-1527,** 1524 (ill.), 1526 (ill.)
Plasma *2:* 326, *7:* 1246
Plasma membrane *3:* 432
Plastic surgery *8:* **1527-1531,** 1530 (ill.), *10:* 1857
Plastics *8:* **1532-1534,** 1532 (ill.)
Plastids *3:* 436
Plate tectonics *8:* **1534-1539,** 1536 (ill.), 1538 (ill.)
Platelets *2:* 329
Platinum *8:* 1566, 1569-1570
Pluto (planet) *8:* **1539-1542,** 1541 (ill.)
Pneumonia *9:* 1681
Pneumonic plague *8:* 1520
Poisons and toxins *8:* **1542-1546,** 1545 (ill.)
Polar and nonpolar bonds *3:* 456
Poliomyelitis *8:* **1546-1549,** 1548 (ill.), *10:* 1958
Pollination *5:* 880-882
Pollution *1:* 9, *8:* **1549-1558,** 1554 (ill.), 1557 (ill.)

Pollution control *8:* **1558-1562,** 1559 (ill.), 1560 (ill.)
Polonium *7:* 1449, 1450
Polygons *8:* **1562-1563,** 1562 (ill.)
Polymers *8:* **1563-1566,** 1565 (ill.)
Polysaccharides *2:* 388
Pompeii *5:* 828, *10:* 1997
Pons, Stanley *7:* 1371
Pontiac Fever *6:* 1181
Pope Gregory XIII *2:* 373
Porpoises *3:* 448
Positive reinforcement. *See* **Reinforcement, positive and negative**
Positron *1:* 163, *4:* 772
Positron-emission tomography. *See* **PET scans**
Positrons *10:* 1832, 1834
Post-it™ notes *1:* 38
Post-traumatic stress disorder *9:* 1826
Potassium *1:* 102
Potassium salts *6:* 1095
Potential difference *4:* 738, 744
Pottery *3:* 447
Praseodymium *6:* 1163
Precambrian era *5:* 988, *8:* 1459
Precious metals *8:* **1566-1570,** 1568 (ill.)
Pregnancy, effect of alcohol on *1:* 87
Pregnancy, Rh factor in *9:* 1684
Pressure *8:* **1570-1571**
Priestley, Joseph *2:* 394, 404, *3:* 525, *7:* 1345, 1444
Primary succession *10:* 1837, 2026
Primates *8:* **1571-1575,** 1573 (ill.), 1574 (ill.)
Probability theory *8:* **1575-1578**
Procaryotae *2:* 253
Progesterone *5:* 800
Projectile motion. *See* **Ballistics**
Prokaryotes *3:* 429
Promethium *6:* 1163
Proof (mathematics) *8:* **1578-1579**
Propanol *1:* 91
Prosthetics *8:* **1579-1583,** 1581 (ill.), 1582 (ill.)
Protease inhibitors *8:* **1583-1586,** 1585 (ill.)
Proteins *7:* 1399, *8:* **1586-1589,** 1588 (ill.)
Protons *10:* 1830, 1832
Protozoa *8:* 1470, **1590-1592,** 1590 (ill.)
Psilocybin *6:* 1029
Psychiatry *8:* **1592-1594**
Psychoanalysis *8:* 1594
Psychology *8:* **1594-1596**
Psychosis *8:* **1596-1598**
Ptolemaic system *3:* 574
Puberty *8:* **1599-1601,** *9:* 1670
Pulley *6:* 1207
Pulsars *7:* 1340
Pyrenees *5:* 826
Pyroclastic flow *10:* 1996
Pyrrophyta *1:* 94
Pythagoras of Samos *8:* 1601
Pythagorean theorem *8:* **1601**
Pytheas *10:* 1890

Q

Qualitative analysis *8:* **1603-1604**
Quantitative analysis *8:* **1604-1607**
Quantum mechanics *8:* **1607-1609**
Quantum number *4:* 772
Quarks *10:* 1830
Quartz *2:* 400
Quasars *8:* **1609-1613,** 1611 (ill.)

R

Rabies *10:* 1958
Radar *8:* **1613-1615,** 1614 (ill.)
Radial keratotomy *8:* **1615-1618,** 1618 (ill.)
Radiation *6:* 1044, *8:* **1619-1621**
Radiation exposure *8:* **1621-1625,** 1623 (ill.), 1625 (ill.)
Radio *8:* **1626-1628,** 1628 (ill.)
Radio astronomy *8:* **1633-1637,** 1635 (ill.)
Radio waves *4:* 765
Radioactive decay dating *4:* 616
Radioactive fallout *7:* 1385, 1386
Radioactive isotopes *6:* 1142, *7:* 1373
Radioactive tracers *8:* **1629-1630**
Radioactivity *8:* **1630-1633**
Radiocarbon dating *1:* 176
Radiology *8:* **1637-1641,** 1640 (ill.)
Radionuclides *7:* 1372
Radiosonde *2:* 216
Radium *1:* 105

Radon *7:* 1349, 1350
Rain forests *2:* 295, *8:* **1641-1645,** 1643 (ill.), 1644 (ill.)
Rainbows *2:* 222
Rainforests *8:* 1643 (ill.)
Ramjets *6:* 1144
Rare earth elements *6:* 1164
Rat-kangaroos *6:* 1157
Rational numbers *1:* 180
Rawinsonde *2:* 216
Reaction, chemical
Reaction, chemical *9:* **1647-1649,** 1649 (ill.)
Reality engine *10:* 1969, 1970
Reber, Grote *8:* 1635
Receptor cells *8:* 1484
Recommended Dietary Allowances *10:* 1984
Reconstructive surgery. *See* **Plastic surgery**
Recycling *9:* **1650-1653,** 1650 (ill.), 1651 (ill.), *10:* 2009
Red algae *1:* 94
Red blood cells *2:* 327, 328 (ill.)
Red giants *9:* **1653-1654**
Red tides *1:* 96
Redox reactions *7:* 1439, 1441
Redshift *8:* 1611, *9:* **1654-1656,** 1656 (ill.)
Reflector telescopes *10:* 1871
Refractor telescopes *10:* 1870
Reines, Frederick *10:* 1833
Reinforcement, positive and negative *9:* **1657-1659**
Reis, Johann Philipp *10:* 1867
Reitz, Bruce *10:* 1926
Relative dating *4:* 616
Relative motion *9:* 1660
Relativity, theory of *9:* **1659-1664**
Relaxation techniques *1:* 118
REM sleep *9:* 1747
Reproduction *9:* **1664-1667,** 1664 (ill.), 1666 (ill.)
Reproductive system *9:* **1667-1670,** 1669 (ill.)
Reptiles *9:* **1670-1672,** 1671 (ill.)
Reptiles, age of *8:* 1462
Respiration
Respiration *2:* 392, *9:* **1672-1677**
Respiratory system *9:* **1677-1683,** 1679 (ill.), 1682 (ill.)
Retroviruses *10:* 1978
Reye's syndrome *1:* 8
Rh factor *9:* **1683-1685,** 1684 (ill.)
Rheumatoid arthritis *1:* 183
Rhinoplasty *8:* 1527
Rhodophyta *1:* 94
Ribonucleic acid *7:* 1390, 1392 (ill.)
Rickets *6:* 1219, *7:* 1403
Riemann, Georg Friedrich Bernhard *10:* 1899
Rift valleys *7:* 1303
Ritalin *2:* 238
Rivers *9:* **1685-1690,** 1687 (ill.), 1689 (ill.)
RNA *7:* 1390, 1392 (ill.)
Robert Fulton *10:* 1835
Robotics *1:* 189, *9:* **1690-1692,** 1691 (ill.)
Robson, Mount *7:* 1357
Rock carvings and paintings *8:* 1491
Rock cycle *9:* 1705
Rockets and missiles *9:* **1693-1701,** 1695 (ill.), 1697 (ill.), 1780 (ill.)
Rocks *9:* **1701-1706,** 1701 (ill.), 1703 (ill.), 1704 (ill.)
Rocky Mountains *7:* 1301, 1357
Roentgen, William *10:* 2033
Rogers, Carl *8:* 1596
Root, Elijah King *7:* 1237
Ross Ice Shelf *1:* 149
Roundworms *8:* 1471
RR Lyrae stars *10:* 1964
RU-486 *3:* 565
Rubidium *1:* 102
Rural techno-ecosystems *2:* 302
Rush, Benjamin *9:* 1713
Rust *7:* 1442
Rutherford, Daniel *7:* 1345
Rutherford, Ernest *2:* 233, *7:* 1337

S

Sabin vaccine *8:* 1548
Sabin, Albert *8:* 1549
Sahara Desert *1:* 52
St. Helens, Mount *10:* 1996
Salicylic acid *1:* 6
Salk vaccine *8:* 1548
Salk, Jonas *8:* 1548, *10:* 1959
Salyut 1 *9:* 1781, 1788

Samarium *6:* 1163
San Andreas Fault *5:* 854
Sandage, Allan *8:* 1610
Sarcodina *8:* 1592
Satellite television *10:* 1877
Satellites *9:* **1707-1708,** 1707 (ill.)
Saturn (planet) *9:* **1708-1712,** 1709 (ill.), 1710 (ill.)
Savanna *2:* 296
Savants *9:* **1712-1715**
Saxitoxin *2:* 288
Scanning Tunneling Microscopy *10:* 1939
Scaphopoda *7:* 1289
Scheele, Carl *7:* 1345, 1444
Scheele, Karl Wilhelm *3:* 525, *6:* 1032
Schiaparelli, Giovanni *7:* 1263
Schizophrenia *8:* 1596, *9:* **1716-1722,** 1718 (ill.), 1721 (ill.)
Schmidt, Maarten *8:* 1611
Scientific method *9:* **1722-1726**
Scorpions *1:* 169
Screw *6:* 1208
Scurvy *6:* 1218, *10:* 1981, 1989
Seamounts *10:* 1994
Seashore biome *2:* 301
Seasons *9:* **1726-1729,** 1726 (ill.)
Second law of motion *6:* 1171, *7:* 1235
Second law of planetary motion *7:* 1426
Second law of thermodynamics *10:* 1886
Secondary cells *2:* 270
Secondary succession *10:* 1837, *10:* 1838
The Secret of Nature Revealed *5:* 877
Sedimentary rocks *9:* 1703
Seeds *9:* **1729-1733,** 1732 (ill.)
Segré, Emilio *1:* 163, *4:* 775, *6:* 1035, *10:* 1913
Seismic waves *4:* 703
Selenium *7:* 1449
Semaphore *10:* 1864
Semiconductors *4:* 666, 734, *10:* 1910, 1910
Semi-evergreen tropical forest *2:* 298
Senility *4:* 622
Senses and perception *8:* 1482
Septicemia plague *8:* 1519
Serotonin *2:* 350
Serpentines *1:* 191
Sertürner, Friedrich *1:* 33
Set theory *9:* **1733-1735,** 1734 (ill.), 1735 (ill.)
Sexual reproduction *9:* 1666
Sexually transmitted diseases *9:* **1735-1739,** 1737 (ill.), 1738 (ill.)
Shell shock *9:* 1826
Shepard, Alan *9:* 1779
Shockley, William *10:* 1910
Shoemaker, Carolyn *6:* 1151
Shoemaker-Levy 9 (comet) *6:* 1151
Shooting stars. *See* **Meteors and meteorites**
Shumway, Norman *10:* 1926
SI system *10:* 1950
Sickle-cell anemia *2:* 320
SIDS. *See* **Sudden infant death syndrome (SIDS)**
Significance of relativity theory *9:* 1663
Silicon *2:* 400, 401
Silicon carbide *1:* 2
Silver *8:* 1566, 1569
Simpson, James Young *1:* 143
Sitter, Willem de *3:* 575
Skeletal muscles *7:* 1310 (ill.), **1311-1313**
Skeletal system *9:* **1739-1743,** 1740 (ill.), 1742 (ill.)
Skin *2:* 362
Skylab *9:* 1781, 1788
Slash-and-burn agriculture *9:* **1743-1744,** 1744 (ill.)
Sleep and sleep disorders *9:* **1745-1749,** 1748 (ill.)
Sleep apnea *9:* 1749, *10:* 1841
Slipher, Vesto Melvin *9:* 1654
Smallpox *10:* 1957
Smell *9:* **1750-1752,** 1750 (ill.)
Smoking *1:* 34, 119, *3:* 476, *9:* 1682
Smoking (food preservation) *5:* 890
Smooth muscles *7:* 1312
Snakes *9:* **1752-1756,** 1754 (ill.)
Soaps and detergents *9:* **1756-1758**
Sobrero, Ascanio *5:* 844
Sodium *1:* 100, 101 (ill.)
Sodium chloride *6:* 1096
Software *3:* 549-554
Soil *9:* **1758-1762.** 1760 (ill.)
Soil conditioners *1:* 67
Solar activity cycle *10:* 1848

Solar cells *8:* 1504, 1505
Solar eclipses *4:* 724
Solar flares *10:* 1846, 1848 (ill.)
Solar power *1:* 115, 115 (ill.)
Solar system *9:* **1762-1767,** 1764 (ill.), 1766 (ill.)
Solstice *9:* 1728
Solution *9:* **1767-1770**
Somatotropic hormone *5:* 797
Sonar *1:* 22, *9:* **1770-1772**
Sørenson, Søren *8:* 1495
Sound. *See* **Acoustics**
South America *9:* **1772-1776,** 1773 (ill.), 1775 (ill.)
South Asia *1:* 197
Southeast Asia *1:* 199
Space *9:* **1776-1777**
Space probes *9:* **1783-1787,** 1785 (ill.), 1786 (ill.)
Space shuttles *9:* 1782 (ill.), 1783
Space station, international *9:* **1788-1792,** 1789 (ill.)
Space stations *9:* 1781
Space, curvature of *3:* 575, *7:* 1428
Space-filling model *7:* 1286, 1286 (ill.)
Space-time continuum *9:* 1777
Spacecraft, manned *9:* **1777-1783,** 1780 (ill.), 1782 (ill.)
Spacecraft, unmanned *9:* 1783
Specific gravity *4:* 625
Specific heat capacity *6:* 1045
Spectrometer *7:* 1239, 1240 (ill.)
Spectroscopes *9:* 1792
Spectroscopy *9:* **1792-1794,** 1792 (ill.)
Spectrum *9:* 1654, **1794-1796**
Speech *9:* **1796-1799**
Speed of light *6:* 1190
Sperm *4:* 785, *5:* 800, *9:* 1667
Spiders *1:* 169
Spina bifida *2:* 321, 321 (ill.)
Split-brain research *2:* 346
Sponges *9:* **1799-1800,** 1800 (ill.)
Sporozoa *8:* 1592
Sprengel, Christian Konrad *5:* 877
Squid *7:* 1289
Staphylococcus *2:* 258, 289
Star clusters *9:* **1808-1810,** 1808 (ill.)
Starburst galaxies *9:* **1806-1808,** 1806 (ill.)
Stars *9:* **1801-1806,** 1803 (ill.), 1804 (ill.)
  binary stars *2:* 276-278
  brown dwarf *2:* 358-359
  magnetic fields *9:* 1820
  variable stars *10:* 1963-1964
  white dwarf *10:* 2027-2028
Static electricity *4:* 742
Stationary fronts *1:* 82
Statistics *9:* **1810-1817**
Staudinger, Hermann *8:* 1565
STDs. *See* **Sexually transmitted diseases**
Steam engines *9:* **1817-1820,** 1819 (ill.)
Steel industry *6:* 1098
Stellar magnetic fields *9:* **1820-1823,** 1822 (ill.)
Sterilization *3:* 565
Stomach ulcers *4:* 656
Stone, Edward *1:* 6
Stonehenge *1:* 173, 172 (ill.)
Stoney, George Johnstone *4:* 771
Storm surges *9:* **1823-1826,** 1825 (ill.)
Storm tide *9:* 1824
Strassmann, Fritz *7:* 1361
Stratosphere *2:* 213
Streptomycin *1:* 155
Stress *9:* **1826-1828**
Strike lines *5:* 988
Stroke *2:* 350, 351
Strontium *1:* 105
Subatomic particles *10:* **1829-1834,** 1833 (ill.)
Submarine canyons *3:* 562
Submarines *10:* **1834-1836,** 1836 (ill.)
Subtropical evergreen forests *5:* 908
Succession *10:* **1837-1840,** 1839 (ill.)
Sudden infant death syndrome (SIDS) *10:* **1840-1844**
Sulfa drugs *1:* 156
Sulfur *6:* 1096, *7:* 1446
Sulfur cycle *7:* 1448, 1448 (ill.)
Sulfuric acid *7:* 1447
Sun *10:* **1844-1849,** 1847 (ill.), 1848 (ill.)
  stellar magnetic field *9:* 1821
Sun dogs *2:* 224
Sunspots *6:* 1077
Super Collider *8:* 1482
Superclusters *9:* 1809
Superconducting Super Collider *10:* 1852

Superconductors *4:* 734, *10:* **1849-1852,** 1851 (ill.)
Supernova *9:* 1654, *10:* **1852-1854,** 1854 (ill.)
Supersonic flight *1:* 43
Surgery *8:* 1527-1531, *10:* **1855-1858,** 1857 (ill.), 1858 (ill.)
Swamps *10:* 2024
Swan, Joseph Wilson *6:* 1088
Symbolic logic *10:* **1859-1860**
Synchrotron *8:* 1481
Synchrotron radiation *10:* 2037
Synthesis *9:* 1648
Syphilis *9:* 1736, 1738 (ill.)
Système International d'Unités *10:* 1950
Szent-Györyi, Albert *6:* 1219

T

Tagliacozzi, Gasparo *8:* 1528
Tapeworms *8:* 1472
Tarsiers *8:* 1572, 1573 (ill.)
Tasmania *2:* 241
Taste *10:* **1861-1863,** 1861 (ill.), 1862 (ill.)
Taste buds *10:* 1861 (ill.), 1862
Tay-Sachs disease *2:* 320
TCDD *1:* 54, *4:* 668
TCP/IP *6:* 1126
Tears *5:* 852
Technetium *4:* 775, *10:* 1913
Telegraph *10:* **1863-1866**
Telephone *10:* **1866-1869,** 1867 (ill.)
Telescope *10:* **1869-1875,** 1872 (ill.), 1874 (ill.)
Television *5:* 871, *10:* **1875-1879**
Tellurium *7:* 1449, 1450
Temperate grassland *2:* 296
Temperate forests *2:* 295, *5:* 909, *8:* 1644
Temperature *6:* 1044, *10:* **1879-1882**
Terbium *6:* 1163
Terrestrial biomes *2:* 293
Testes *5:* 800, *8:* 1599, *9:* 1667
Testosterone *8:* 1599
Tetanus *2:* 258
Tetracyclines *1:* 158
Tetrahydrocannabinol *6:* 1224
Textile industry *6:* 1097
Thalamus *2:* 342
Thallium *1:* 126
THC *6:* 1224
Therapy, physical *8:* 1511-1513
Thermal energy *6:* 1044
Thermal expansion *5:* 842-843, *10:* **1883-1884,** 1883 (ill.)
Thermodynamics *10:* **1885-1887**
Thermoluminescence *4:* 618
Thermometers *10:* 1881
Thermonuclear reactions *7:* 1368
Thermoplastic *8:* 1533
Thermosetting plastics *8:* 1533
Thermosphere *2:* 213
Thiamine. *See* **Vitamin B1**
Third law of motion *6:* 1171
Third law of planetary motion *7:* 1426
Thomson, Benjamin *10:* 1885
Thomson, J. J. *2:* 233, *4:* 771
Thomson, William *10:* 1885, 1882
Thorium *1:* 26
Thulium *6:* 1163
Thunder *10:* 1889
Thunderstorms *10:* **1887-1890,** 1889 (ill.)
Thymus *2:* 329, *5:* 798
Thyroxine *6:* 1035
Ticks *1:* 170, *8:* 1475
Tidal and ocean thermal energy *1:* 117
Tides *1:* 117, *10:* **1890-1894,** 1892 (ill.), 1893 (ill.)
Tigers *5:* 859
Time *10:* **1894-1897,** 1896 (ill.)
Tin *2:* 401, 402
TIROS 1 *2:* 217
Titan *9:* 1711
Titania *10:* 1954
Titanic *6:* 1081
Titius, Johann *1:* 201
Tools, hand *6:* 1036
Topology *10:* **1897-1899,** 1898 (ill.), 1899 (ill.)
Tornadoes *10:* **1900-1903,** 1900 (ill.)
Torricelli, Evangelista *2:* 265
Touch *10:* **1903-1905**
Toxins, poisons and *8:* 1542-1546
Tranquilizers *10:* **1905-1908,** 1907 (ill.)
Transformers *10:* **1908-1910,** 1909 (ill.)
Transistors *10:* 1962, **1910-1913,** 1912 (ill.)

Transition elements *10:* **1913-1923,** 1917 (ill.), 1920 (ill.), 1922 (ill.)
Transplants, surgical *10:* **1923-1927,** 1926 (ill.)
Transuranium elements *1:* 24
Transverse wave *10:* 2015
Tree-ring dating *4:* 619
Trees *10:* **1927-1931,** 1928 (ill.)
Trematodes *8:* 1473
Trenches, ocean *7:* 1410
Trevithick, Richard *6:* 1099
Trichomoniasis *9:* 1735
Trigonometric functions *10:* 1931
Trigonometry *10:* **1931-1933**
Triode *10:* 1961
Triton *7:* 1332
Tropical evergreen forests *5:* 908
Tropical grasslands *2:* 296
Tropical rain forests *5:* 908, *8:* 1642
Tropism *2:* 271
Troposphere *2:* 212
Trusses *2:* 356
Ts'ai Lun *8:* 1463
Tularemia *2:* 289
Tumors *10:* **1934-1937,** 1934 (ill.), 1936 (ill.)
Tundra *2:* 293
Tunneling *10:* **1937-1939,** 1937 (ill.)
Turbojets *6:* 1146
Turboprop engines *6:* 1146
Turbulent flow *1:* 40

U

U.S.S. *Nautilus* 10: *1836*
Ulcers (stomach) *4:* 656
Ultrasonics *1:* 23, *10:* **1941-1943,** 1942 (ill.)
Ultrasound *8:* 1640
Ultraviolet astronomy *10:* **1943-1946,** 1945 (ill.)
Ultraviolet radiation *4:* 765
Ultraviolet telescopes *10:* 1945
Uluru *2:* 240
Umbriel *10:* 1954
Uncertainty principle *8:* 1609
Uniformitarianism *10:* **1946-1947**
Units and standards *7:* 1265, *10:* **1948-1952**
Universe, creation of *2:* 273
Uranium *1:* 25, *7:* 1361, 1363
Uranus (planet) *10:* **1952-1955,** 1953 (ill.), 1954 (ill.)
Urban-Industrial techno-ecosystems *2:* 302
Urea *4:* 645
Urethra *5:* 841
Urine *1:* 139, *5:* 840
Urodeles *1:* 137
Ussher, James *10:* 1946

V

Vaccination.*See* **Immunization**
Vaccines *10:* **1957-1960,** 1959 (ill.)
Vacuoles *3:* 436
Vacuum *10:* **1960-1961**
Vacuum tube diode *4:* 666
Vacuum tubes *3:* 416, *10:* **1961-1963**
Vail, Alfred *10:* 1865
Van de Graaff *4:* 742 (ill.), *8:* 1475
Van de Graaff, Robert Jemison *8:* 1475
Van Helmont, Jan Baptista *2:* 337, 393, 404
Variable stars *10:* **1963-1964**
Vasectomy *3:* 565
Venereal disease *9:* 1735
Venter, J. Craig *6:* 1063
Venus (planet) *10:* **1964-1967,** 1965 (ill.), 1966 (ill.)
Vertebrates *10:* **1967-1968,** 1967 (ill.)
Vesalius, Andreas *1:* 139
Vesicles *3:* 433
Vibrations, infrasonic *1:* 18
Video disk recording *10:* 1969
Video recording *10:* **1968-1969**
Vidie, Lucien *2:* 266
Viè, Françoise *1:* 97
Vietnam War *1:* 55, *3:* 460
Virtual reality *10:* **1969-1974,** 1973 (ill.)
Viruses *10:* **1974-1981,** 1976 (ill.), 1979 (ill.)
Visible spectrum *2:* 221
Visualization *1:* 119
Vitamin A *6:* 1220, *10:* 1984
Vitamin B *10:* 1986
Vitamin B_1 *6:* 1219
Vitamin B_3 *6:* 1219
Vitamin C *6:* 1219, *10:* 1981, 1987, 1988 (ill.)

Vitamin D *6:* 1219, *10:* 1985
Vitamin E *10:* 1985
Vitamin K *10:* 1986
Vitamins *7:* 1401, *10:* **1981-1989,** 1988 (ill.)
Vitreous humor *5:* 851
Viviparous animals *2:* 317
Vivisection *10:* **1989-1992**
Volcanoes *7:* 1411, *10:* **1992-1999,** 1997 (ill.), 1998 (ill.)
Volta, Alessandro *4:* 752, *10:* 1865
Voltaic cells *3:* 437
Volume *10:* **1999-2002**
Von Graefe, Karl Ferdinand *8:* 1527
Vostok *9:* 1778
Voyager 2 *10:* 1953
Vrba, Elisabeth *1:* 32

W

Waksman, Selman *1:* 157
Wallabies, kangaroos and *6:* 1153-1157
Wallace, Alfred Russell *5:* 834
War, Peter *10:* 1924
Warfare, biological. *2:* 287-290
Warm fronts *1:* 82
Waste management *7:* 1379, *10:* **2003-2010,** 2005 (ill.), 2006 (ill.), 2008 (ill.)
Water *10:* **2010-2014,** 2013 (ill.)
Water cycle. *See* **Hydrologic cycle**
Water pollution *8:* 1556, 1561
Watson, James *3:* 473, *4:* 786, *5:* 973, 980 (ill.), 982, *7:* 1389
Watson, John B. *8:* 1595
Watt *4:* 746
Watt, James *3:* 606, *9:* 1818
Wave motion *10:* **2014-2017**
Wave theory of light *6:* 1187
Wavelength *4:* 763
Waxes *6:* 1191
Weather *3:* 608-610, *10:* 1887-1890, 1900-1903, **2017-2020,** 2017 (ill.)
Weather balloons *2:* 216 (ill.)
Weather forecasting *10:* **2020-2023,** 2021 (ill.), 2023 (ill.)
Weather, effect of El Niño on *4:* 782
Wedge *6:* 1207
Wegener, Alfred *8:* 1534
Weights and measures. *See* **Units and standards**
Welding *4:* 736
Well, Percival *8:* 1539
Wells, Horace *1:* 142
Went, Frits *6:* 1051
Wertheimer, Max *8:* 1595
Wetlands *2:* 299
Wetlands *10:* **2024-2027,** 2024 (ill.)
Whales *3:* 448
Wheatstone, Charles *10:* 1865
Wheel *6:* 1207
White blood cells *2:* 328, 1085 (ill.)
White dwarf *10:* **2027-2028,** 2027 (ill.)
Whitney, Eli *6:* 1098, *7:* 1237
Whole numbers *1:* 180
Wiles, Andrew J. *7:* 1394
Willis, Thomas *4:* 640, *9:* 1718
Wilmut, Ian *3:* 487
Wilson, Robert *8:* 1637
Wind *10:* **2028-2031,** 2030 (ill.)
Wind cells *2:* 218
Wind power *1:* 114, 114 (ill.)
Wind shear *10:* 2031
Withdrawal *1:* 35
Wöhler, Friedrich *7:* 1428
Wöhler, Hans *1:* 124
Wolves *2:* 383, 383 (ill.)
World Wide Web *6:* 1128
WORMs *3:* 533
Wright, Orville *1:* 75, 77
Wright, Wilbur *1:* 77
Wundt, Wilhelm *8:* 1594

X

X rays *4:* 764, *8:* 1639, *10:* 1855, **2033-2038,** 2035 (ill.), 2036 (ill.)
X-ray astronomy *10:* **2038-2041,** 2040 (ill.)
X-ray diffraction *4:* 650
Xanthophyta *1:* 95
Xenon *7:* 1349, 1352
Xerography *8:* 1502
Xerophthalmia *6:* 1220
Xylem *6:* 1175, *8:* 1523

Y

Yangtze River *1:* 199
Yeast *10:* **2043-2045,** 2044 (ill.)
Yellow-green algae *1:* 95

Yoga *1:* 119
Young, Thomas *6:* 1113
Ytterbium *6:* 1163

Z

Zeeman effect *9:* 1823
Zeeman-Doppler imaging *9:* 1823
Zehnder, L. *6:* 1116
Zeppelin, Ferdinand von *1:* 75
Zero *10:* **2047-2048**
Zoophobia *8:* 1497
Zooplankton *8:* 1521, 1522
Zosimos of Panopolis *1:* 84
Zweig, George *10:* 1829
Zworykin, Vladimir *10:* 1875
Zygote *4:* 787